Metallurgy of basic weld metal

Metallurgy of basic weld metal

G M EVANS AND N BAILEY

WILLIAM ANDREW INC.

ABINGTON PUBLISHING

Woodhead Publishing Ltd in association with The Welding Institute

Cambridge England

Published by Abington Publishing
Woodhead Publishing Limited, Abington Hall,
Abington, Cambridge CB1 6AH, England

Published in North and South America by William Andrew Inc.,
13 Eaton Avenue, Norwich, NY 13815, USA

First published 1997 Abington Publishing and William Andrew Inc.

British Library Cataloguing in Publication Data
A catalogue record for this book is available from the British Library.

ISBN 1 85573 243 2 (Woodhead Publishing Limited)
ISBN 1 884207 57 X (William Andrew Inc.)

Designed by Geoff Green
Typeset by Best-set Typesetter Limited, Hong Kong
Printed in Great Britain by St Edmundsbury Press Limited, Bury St Edmunds, Suffolk

Contents

Part IV Microalloying of C–Mn steel weld metals

Part V Microalloying of high purity low alloy steel weld metals

Part VI Metallography

Acknowledgements

The authors are grateful to the Welding Division of the Air Liquide Group for permission to publish this compendium. They are also indebted to the staff of the research department of Oerlikon-Welding Ltd (Zürich), in particular to Dr H Baach who instigated and supervised the work, to Dr P Korber and Mr HH Wolfensburger for the precise chemical analysis, to Mr R Züllig for metallographic assistance and to Mr V Musarra for so skillfully welding the test plates over a 20 year period.

Thanks are also extended to the management of TWI (The Welding Institute) for allowing secondment of one of the authors (GME) and providing assistance in preparation of the manuscript and in particular to the staff: Dr RE Dolby, Dr DJ Abson, Mr PHM Hart, Mr RJ Pargeter; and former staff Dr DJ Widgery, Dr AR Jones and Dr MT Harvey for their help.

We also thank the following colleagues for their invaluable contributions to this work:

Prof RC Cochrane, formerly BSC plc, now with Leeds University, UK;
Prof JMA Rebello, Federal University of Rio de Janeiro, Brasil;
Dr MQ Johnson (now with EWI) and Prof GR Edwards, Colorado School of Mines, USA;
Prof IK Pokhodnya, EO Paton Welding Institute, Kiev, Ukraine;
Prof G l'Espérance, Dr S St Laurent and Mr C Blais, École Polytechnique, Montreal, Canada;
Dr M Koçak, Dr M Es-Souni and Dr P Beaven, GKSS-Forschungszentrum, Geesthacht, Germany;
Dr M Dadian and the late Mr HP Granjon, Institut de Soudure, Paris, France;
The late Prof N Christensen, Prof Ø Grong and Dr AO Kluken, SINTEF, Trondheim, Norway;

Dr J Bosansky, Welding Research Institute, Bratislava, Slovakia;
Dr JA Whiteman and Dr AR Mills, Sheffield University, UK;
Dr JG Garland and Dr DS Taylor, Oerlikon Welding Ltd, Hayes, UK;
Prof G den Ouden, Delft University of Technology, Netherlands;
Prof HH Cerjak and Mr E Letofsky, Technical University of Graz, Austria.

Finally, we thank Dr T Boniszewski (Consultant) for helpful discussions and encouragement during the course of the studies described in this monograph.

1

Introduction

Manual metal arc (MMA), or shielded metal arc welding (SMAW) as it is known in the USA and Canada, was discovered near the beginning of the twentieth century and during most of that time it has been threatened by newer, more sophisticated and more efficient welding processes. Nevertheless, it still survives – particularly where high quality is required.

Five factors contribute to this success. Firstly, MMA welding is controlled by a man and not a robot so that it can be used to overcome departures from the ideal in the joint preparation. It can be used where the root gap is large or variable or where there is appreciable misalignment between the parts being joined. Secondly, it can be used outdoors where shielding from winds is minimal. Thirdly, welds can be made in restricted spaces where other processes cannot easily be deployed. Fourthly, electrodes of special composition can be made inexpensively in small quantities. In consequence, weld metals have been developed having combinations of properties, particularly strength and toughness which, in most cases, are the equal of anything developed for the newer welding processes. Finally, the equipment for MMA welding is of low capital cost and is compact – needing only a lead from the power source to the electrode holder. The major disadvantages of the process are the high degree of skill required of the welder and a relatively low productivity in terms of rate of metal deposition.

This monograph describes a systematic study of the mechanical properties and microstructures of ferritic steel manual metal arc weld metals. The work described was commenced over 20 years ago, before there was a clear understanding of the part played by elements in very small quantities, or even before routine techniques for their analysis were developed. The earlier part of the book describes, in effect, the development of an understanding of the role of the principal elements and impurities which were commonly analysed for at the time in mild steels and C–Mn and low alloy steel weld metals, using then standard commercial raw materials. These inevitably led to low, but significant, levels of the micro-alloying elements titanium, niobium and vanadium in the weld metal. Later chapters describe effects of these micro-alloying elements in a basic weld metal essentially free from those elements, and also of the major alloying elements which had been originally studied, but in much purer weld metal. In some cases the latter had surprisingly different properties from the same nominal compositions studied in the earlier chapters.

Although the book shows how the results relate to existing theories of weld metal mechanical properties and microstructures, it does not attempt to construct new models. It does, however, provide a large database which will be of use to researchers trying to develop detailed systems for understanding and controlling weld metal properties.

The MMA (SMAW) welding process

Welding processes are divided into those which join the parent metals without adding filler – the autogenous processes and those which, like MMA, require a filler. Tungsten inert gas (TIG or gas tungsten arc – GTAW) welding spans the two and can be used with or without a filler, so that dilution of parent metal in the weld can vary from 100% to very little. The welding processes with filler can be divided into those, such as metal inert gas or metal active gas (MIG or MAG, i.e. gas metal arc welding – GMAW) which use a (normally) solid wire with a shielding gas and those like MMA and submerged-arc (SAW) which use a wire and flux system. In autogenous welds (otherwise outside the scope of this book), the weld metal is, except in special circumstances, similar to a heat-affected zone (HAZ), except that it has been melted and solidified without the benefits of any working and heat treatment that the parent metal has received.

A weld with filler has a composition intermediate between the parent metal and the welding wire but modified by the result of chemical reactions between the weld metal and shielding gas and/or flux. In MMA welding of ferritic steels, shielding gas (usually CO from carbonates in the electrode coating) adds to any effect of the flux. Although several types of MMA electrode are used, the present work used only basic (formerly known as low hydrogen) electrodes as they, rather than rutile or cellulosic electrodes, are capable of giving good toughness, particularly when higher strength is needed.

Factors which control the properties of welds are:

(i)   cooling rate or speed (which in ferritic steel welds controls the transformation products as well as the initial cast structure);

(ii)   composition;

(iii)   overall structure of the weld, as defined by the number of passes used to make it and how each pass alters the previously deposited weld metal.

Cooling speed (usually described in terms of the time to cool from 800 °C to 500 °C, or $T_{8/5}$) is controlled by the heat input (or arc energy, expressed as kJ/mm of weld length), by the temperature at which welding starts (i.e. the preheat or, in the case of all except the first run of a multipass weld, the interpass temperature) and by the thickness of the parts being joined (the combined thickness). Fast cooling leads to a finer macrostructure, usually to a less coarse microstructure, higher hardness and greater strength. The effect of increasing cooling rate on toughness in ferritic steels is very variable. It is complicated by the different microstructures which result from different cooling rates and their coarsening at lower cooling rates, which tends to reduce toughness. Although most welding processes can be used over a wide range of heat input, the common welding processes over their more common ranges of use can be ranked in order of increasing heat input as TIG, MMA, self-shielded cored wire, MIG/CO_2 (with either solid or cored wire), submerged-arc and electroslag (including consumable guide).

Mild, C–Mn and alloyed weld metal compositions were originally expressed, like the compositions of mild steels, in terms of the contents of the elements C, Mn, Si, S and P, plus any deliberate alloying elements. The first three elements are both alloy (i.e. strengthening) elements and deoxidisers and the last two are common impurities. Over the years this scheme has become increasingly inadequate and now the elements oxygen, nitrogen, titanium and aluminium, as well as vanadium, niobium and boron are all important and should be reported.

Oxygen content gives a useful means of classifying weld metals with respect to the type of flux/shielding system used, and the state of deoxidation of the weld metal and its likely inclusion content. Typical contents vary from <0.01% for TIG welds, 0.02–0.04% for basic submerged-arc, 0.03–0.05% for basic MMA, rather higher (up to 0.1%) for some other types of MMA and most MIG and CO_2 and rising to 0.12% for submerged-arc with manganese silicate fluxes. Nitrogen content is a guide to the efficiency of shielding from the air during welding (provided the contents of the wire and parent steels are known), whilst the other elements, as will be shown later, are important in controlling the microstructure and properties.

The arrangement of the weld passes is important in ferritic steel welds because each pass heats some of the existing weld metal above its transformation temperature to form austenite. Heated above a certain higher temperature, the austenite grains grow and give rise, on cooling and transformation, to a coarse-grained (or high temperature) HAZ, whilst that heated to lower temperatures within the austenite range results in a fine-grained (or low temperature) HAZ. The microstructure of the coarse-grained HAZ resembles that of the as-deposited weld metal. That of the fine-grained HAZ is based on fine equiaxed ferrite and is generally desirable, as it usually exhibits the finest ferrite grain size and the best toughness. In extreme cases it is possible to devise a welding procedure which, after the top bead has been ground off, gives a completely fine-grained and correspondingly tough deposit. However, in most practical situations, the presence of microstructures which are as-deposited (or only slightly altered from that condition) and coarse reheated must be accepted.

Mechanical properties and microstructure

In most metallurgical systems, an increase in strength is accompanied by a corresponding fall in toughness. Only if the increasing strength is accompanied by refinement of the microstructure – i.e. a decrease in the effective (ferrite) grain size – can the two properties be increased simultaneously.

Such refinement in ferritic steel weld metals can be achieved by alloying in such a way as to change the microstructure from one of coarse ferrite to a microstructure peculiar to ferritic steel weld metals – acicular ferrite. Acicular ferrite is of similar ferrite grain size to the fine ferrite formed in multipass weld metal when the microstructure of the underlying weld runs is refined by reheating within the lower temperature austenite range by succeeding passes.

In welding ferritic steels for structural use, achieving sufficient strength to match that of the parent steel is not a problem, unless the steel is of extremely high strength, i.e. with a yield strength exceeding about $800 \, \text{N/mm}^2$. The major problem is to obtain adequate toughness in the weld metal – particularly when the joint cannot be given a stress relief or post-weld heat treatment (PWHT). Post-weld normalising or quenching and tempering are not normally options, partly because of the physical difficulties (avoiding distortion and oxidation) and partly because weld metals of low carbon content (needed to minimise potential cracking problems) lose a large proportion of their strength on such treatment.[1]

Toughness is a major problem because, in the as-welded condition (in which most welds enter service), tensile residual stresses up to the magnitude of the yield strength of the deposited weld metal are likely to be present at or near one or other surfaces of the weldment. Such high stresses require a correspondingly high level of toughness in the weld metal to achieve an acceptably low risk of brittle fracture in service.

Principal aims

The aims of the project described in this book were:

(i)   to discover how compositional variations influence the mechanical properties and microstructure of basic mild steel weld metals with less than 5% alloying elements;
(ii)   from the results of the experimental programme, to improve existing MMA electrodes for high quality welding of mild, C–Mn, low alloy, Cr–Mo and medium alloy steels with yield strengths up to about 700 N/mm^2;
(iii)   to relate the results to current ideas about the relationships between weld metal composition, microstructure, strength and toughness;
(iv)   to provide a database for modelling purposes to be able to predict weld metal properties.

Experimental organisation

Most of the work was carried out in the research laboratories of Oerlikon Welding Ltd, Zurich, a member of the Air Liquide Welding Group, although some small parts of the work (e.g. particular metallographic examinations) were often sub-contracted. Besides the principal reference in the open literature, the reports were usually published in the Oerlikon house journal *Schweissmitteilungen* and released as Documents to the International Institute of Welding (IIW/IIS); references to these sources have been added to the open literature references (which are mainly in the *Welding Journal*).

Other parts of the work were carried out as joint projects between Oerlikon and various co-operating organisations. The establishments which collaborated in the project were BSC (British Steel Corporation, later British Steel, UK), CSM (Colorado School of Mines, USA), DUT (Delft University of Technology, Netherlands), EO Paton Electric Welding Institute (Kiev, Ukraine), EPM (École Polytechnique de Montréal, Canada), GKSS Forschungszentrum Geesthacht GmbH (Germany), IS (French Welding Institute, Paris), Oerlikon Welding Ltd (UK), Sheffield University (UK), SINTEF (Foundation for Scientific and Industrial Research at the Technical

University of Norway), TUG (Technical University of Graz, Austria), TWI (The Welding Institute, UK), UFRJ (Federal University of Rio de Janiero, Brasil), and the Welding Research Institute (Bratislava, Slovakia).

Grateful thanks are given for the help and advice provided by these organisations and their staff. This international co-operation was greatly aided by discussions within the IIW, particularly in Commission II, of which one of the authors (GME) has long been a member and chairman of Sub-commission IIA (The Metallurgy of Weld Metal).

Background

A few years before the start of this work, a paper had been published[1] which examined why the strength of a mild steel weld metal was so much greater than would be expected from a normal steel, cast or wrought, of similar composition and grain size. The authors concluded that a fine, dense dislocation network, formed as the weld cooled under considerable restraint, was responsible for this remarkable strength. The dislocation network was very resistant to temperatures up to 650 or even 700 °C, as used for PWHT or stress relief, although it was broken down at the higher temperatures used for normalising.

Recent literature surveys have been compiled in the UK[2] and Australia,[3] so that only a very brief summary of the state of the art at an early stage of the project is detailed here. A number of works dealing with the subject have, however, been published in recent years.[4-13] Also, there are many technical papers on the development, microstructure and mechanical properties of MMA electrodes and weld metals, whilst the importance of acicular ferrite in achieving good toughness is well established. In an as-welded deposit, alloying with an element which promotes acicular ferrite gives the double benefit of increasing strength and toughness – one of the few instances where such a double benefit is possible without recourse to heat treatment and/or mechanical working.

It is now known[2-13] that to achieve an acicular ferrite microstructure, at least two conditions are necessary:

Firstly, the non-metallic inclusions in the weld metal should have a suitable surface layer – probably containing particles rich in titanium. In practice, this requires a certain level of oxygen in the weld metal and a ratio (in weight per cent) of aluminium (when present) to oxygen below about $9:8$ (the ratio of the weights of the two elements in alumina – Al_2O_3) to allow titanium in the weld to be effective.

Secondly, there should be sufficient alloying elements to replace coarse primary (grain boundary) ferrite by acicular ferrite, but not to such an extent that lower temperature transformation products, such as bainite or martensite, can form.

It is likely that other conditions are necessary, particularly as regards the limits of, and the optimum, titanium content, and the possibility that acicular ferrite can be induced in other circumstances. However, the conditions given above have been found to provide a useful rough, practical guide and to explain several cases where acicular ferrite was unexpectedly absent in practical welding.

At the start of the present project, the role of titanium was not appreciated, although it was known that acicular ferrite could be formed most easily in welds where the weld metal oxygen content lay somewhere between about 0.02 and 0.055%, i.e. a range which includes basic MMA weld metal, as well as basic and semi-basic submerged-arc, and some MIG/MAG deposits made with low oxygen shielding gases. However, self-shielded deposits, although of very low oxygen content, contained little acicular ferrite and this was believed to be associated in some way with their very high aluminium contents.

Because the importance of titanium (and other microalloying elements) only became apparent part way through the project, the earlier results from mild steel and C–Mn weld metals of what was, at the time, normal quality are given in Part I of this monograph. The later results, in which C–Mn welds with low contents and controlled additions of microalloying elements were studied, are presented in Parts III and IV. The influences of major alloying elements are described in Part II with old standard materials and in Part V with controlled microalloy additions. Part VI summarises observations on metallographic features such as inclusions, dislocations and segregation.

References

1 Wheatley JM and Baker RG: 'Factors governing the yield strength of a mild steel weld metal deposited by the metal arc process'. *Brit Weld J* 1963 **10** 23–8.

2 Abson DJ and Pargeter RJ: 'Factors influencing deposited strength, microstructure and toughness of manual metal arc welds suitable for C–Mn steels'. *International Metals Reviews* 1986 **31**(4) 141–94.

3 Schumann O, Powell G and French I: 'A literature review on control of microstructure in gas metal arc (GMA) and flux-cored arc (FCA) welding processes'. WTIA/CRC No 1, Nov 1994, Welding Technology Inst of Australia, Silverwater, NSW.

4 Bailey N: 'Weldability of ferritic steels'. Abington Publishing, 1994.

5 Boniszewski T: 'MMA welding'. *The Metallurgist and Materials Technologist* 1979 11(10) 567–74, (11) 640–3, (12) 697–705.
6 Svenssen LE: 'Control of microstructures and properties in steel arc welds'. CRC Press, 1994.
7 Grong Ø: 'Metallurgical modelling of welding'. The Inst of Materials, London, 1994.
8 Pokhodnya IK (ed): 'Metallurgy of arc welding'. Paton Institute, 1991.
9 Boniszewski T: 'Self-shielded arc welding'. Abington Publishing, Cambridge, 1992.
10 Widgery DJ: 'Tubular wire welding'. Abington Publishing, Cambridge, 1992.
11 Linnert GE: 'Welding metallurgy: carbon and alloy steels'. AWS, 4th edition, 1994.
12 Granjon H: 'Fundamentals of welding metallurgy'. Abington Publishing, Cambridge, 1991.
13 Bhadeshia HKDH: 'Bainite in steels'. The Institute of Materials, 1992.

2

Experimental techniques

This chapter describes the general experimental methods used throughout the work; exceptions are described in the appropriate chapter or section. Such exceptions were necessary when, for example, some of the process parameters themselves were being examined, as in Chapter 4, and when co-operation with a specialist research establishment was sought to investigate more deeply a particular aspect of a problem. Similarly, changes were made when consumables for a particular steel were examined, or when welding or heat treatment conditions were altered to suit a special requirement. Recoveries of elements added during the programme are given later in this chapter to avoid unnecessary detail in later sections.

Electrode production

The all-positional electrodes used were produced using standard techniques for basic (low hydrogen) 25% iron powder covered electrodes (similar to the classification E7018 in the American Welding Society Standard[1]) and were specially formulated to keep all elements as constant as practicable, except for the one under investigation. This, for example, involved reducing the level of ferro-silicon in the coating when other deoxidants were added or increased in quantity. However, in some cases it was not possible to maintain all elements constant without taking unusual steps, which were not attempted; resultant small systematic variations were accepted in such cases.

For the electrodes in Parts I and II, the raw materials used (particularly the type of iron powder) were such that suitable alloying (especially with manganese) was adequate in itself to allow the formation of acicular ferrite. Raw materials for electrodes for Part III onwards were selected so that the content of each of the microalloying elements Al, B, Nb, V and Ti could be kept below 5 ppm in the resultant weld metal if required.

The core wire for the electrodes originated from different batches of rimming steel but was all of a similar composition; a typical analysis, in wt%, was 0.07 C, 0.50 Mn, 0.008 Si, 0.006 S, 0.008 P, 0.02 Cr, 0.003 Mo, 0.03 Ni, 0.02 Cu, 0.0004 Ti, 0.0015 Al, 0.0005 Nb, 0.0005 V, 0.0002 B, 0.020 O and 0.0025 N.

Minerals were dry mixed before adding the tertiary silicate binder to produce a single coating, which was extruded on to 4 mm diameter rimming steel core wires, generally with a coating factor (the ratio of electrode to core wire diameter) of 1.68. The extruded electrodes were baked for 1 hour at 400 °C to achieve a very low weld hydrogen content of ~2.3 mlH/100 g deposited metal according to the appropriate International Standard.[2]

Welding, sectioning, heat treatment and testing

The welds were deposited according to the method in ISO 2560:1973*.[3] The technique is used to qualify manual electrodes and, in the present work, consisted of depositing a weld three runs wide (Fig. 2.1) so that the subsequent mechanical property tests (Fig. 2.2) were carried out on the central run(s). This procedure minimised any effects of possible variations in base plate composition and ensured that the tests were made on deposits which contained a proportion of columnar and high temperature reheated weld metal.*

Two mild steel plates, 20 mm thickness, were tacked to a backing strip and clamped in a jig. Welding was carried out manually using a stringer bead technique in the flat (downhand) position; a welding current of 170 A was generally used, together with an arc voltage of 21 V, DC electrode positive. The welding speed was adjusted to give a heat input (arc energy) of 1 kJ/mm and, after the first series of tests in Chapter 3 (which must in many respects be regarded as preliminary experiments), the interpass temperature was maintained at 200 °C. A single welder carried out all the welding in the programme. Any exceptions to this procedure are clearly stated in the text.

After completion of welding, test panels were sectioned, essentially as shown in Fig. 2.2; the portion used for tensile testing in the

---

* The later versions of ISO 2560 and the American standard AWS A5.1 (revisions of 1973 and 1974, respectively) were brought more into line with each other. In consequence it is no longer allowable with in ISO 2560 to deposit welds three beads wide in order to introduce a realistic amount of columnar and coarse-grained high temperature reheated weld metal into the region subjected to mechanical property testing. For further discussion of this topic see the last section of Chapter 4.

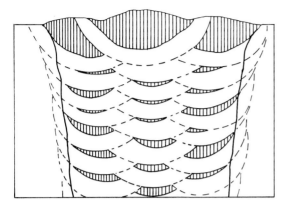

2.1 ISO 2560 deposit showing disposition of weld runs.

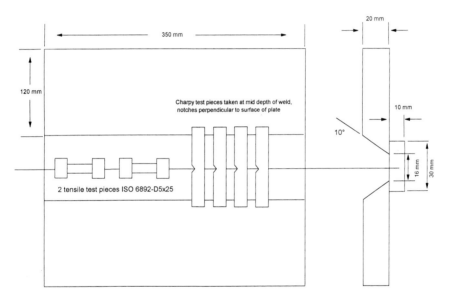

2.2 Part of length of ISO 2560–1973 test assembly showing extraction of testpieces.

'as-welded' condition was heated for 14 hours at 250 °C to remove diffusible hydrogen. This heating is standard when tensile tests are carried out on electrode qualification testpieces,[3] as hydrogen is known to reduce ductility and can give markings known as fisheyes on the fracture surfaces. It does not reduce toughness in tests carried out at high strain rates, such as Charpy tests, and these tests were carried out without such heating.

Hydrogen is removed during qualification testing because it slowly diffuses out of welds at ambient temperature. It is, therefore, of little consequence to the properties of a weld in a fabrication not likely to be used soon after welding. However, in a qualification test, consistent results are required – often in a hurry – so that removal of hydrogen is an essential part of the procedure, as variations in the time between welding and testing are likely to affect results inconsistently.

For metallographic examination, a transverse section of each deposit was examined. After visual examination, more detailed microscopic examination was carried out along the mid-plane of the deposit, concentrating particularly on the top (final and therefore un-reheated) bead and the underlying central bead which had been reheated by the top bead. Quantitative measurements were made to establish prior austenite grain sizes and the proportions of different regions. The point-counting technique developed by Davey and Widgery[4] and adopted by the International Institute of Welding (IIW)[5,6] was employed to determine the proportions of constituents in the final, unreheated bead. The weld metal, after polishing and etching in nital, was examined at a magnification of at least ×500 and traversed in a pre-determined manner and the constituent under the microscope cross-wires at each step was identified and recorded. In some instances, scanning and/or transmission electron microscopy (SEM or TEM) were used to examine ferrite grain sizes, inclusions and fine details of some constituents. Other techniques, including micro-analysis, are described in the appropriate sections.

Nomenclature of constituents of the as-deposited region has varied over the years and has been the subject of much controversy. In this work, the historical developments which have taken place in IIW Sub-Commission IXJ have been followed.

As ferritic weld metal cools, it transforms from austenite to various mixtures of ferrite and cementite as follows:

Primary ferrite – PF – (sometimes divided into grain boundary ferrite and polygonal ferrite);

Ferrite with second phase – FS – (sometimes divided into ferrite with aligned second phase – FSA – and ferrite with unaligned second phase; this constituent has also been known as ferrite side plates or ferrite with aligned MAC [martensite/austenite/carbide] and has been regarded by some as a type of bainite);

Acicular ferrite – AF – (fine grained ferrite plates, nucleated within prior austenite grains – a constituent peculiar to certain ferritic weld metals);

Ferrite-carbide aggregate – FCA – (an intimate mixture of ferrite and cementite);

Bainite – B – (this is properly recognised only in alloyed weld metals, particularly those containing chromium);

Martensite – M – (sometimes divided into twinned martensite and lath martensite);

Carbide – (usually cementite, Fe_3C, in C–Mn and similar weld metals).

The last three constituents cannot be confidently identified by optical (light) microscopy alone. Use of electron microscopy also permits identification of the so-called microphases. These are constituents which occur at the corners between plates of various types of ferrite (particularly FS) and which have been late to transform or have not completely transformed. They are mixtures of retained austenite (A), martensite and carbide. On stress relief, the retained austenite transforms, usually to martensite, and the original martensite tempers to ferrite and carbide.

Hardness and micro-hardness measurements were made on specimens prepared for metallographic examination using the Vickers testing method. Although the results in the text are quoted as 'HV5', the older designation 'DPN' is employed in most of the figures. Similarly 'UTS' is likewise used for 'tensile strength'.

Tests of mechanical properties were carried out in the as-welded condition or after stress relief or PWHT for 2 hours at 580 °C. For tensile tests, sub-standard specimens of the Minitrac type,[7] with a 50 mm gauge length parallel to the welding direction and 5 mm gauge diameter, were tested in duplicate.

Standard Charpy V testpieces were used with their axes perpendicular to the welding direction and their notches perpendicular to the plate surface. The mean 100 J and 28 J (transition) temperatures were determined from the transition curves, the 100 J temperature being considered the more significant. The change in toughness after stress-relief heat treatment was described as a lateral shift, a positive value indicating an increase in transition temperature. To examine the influence of strain ageing, some blanks, from which Charpy specimens were subsequently extracted, were compressed 5 or 10% and aged for 0.5 hour at 250 °C and compared with the as-welded results. Sufficient tests (usually 36) were carried out to establish a transition curve in each condition tested.

Chemical analyses were carried out on used tensile test specimens from each deposit. The analysis technique, appropriate to the level of element being studied, was based on automatic spectrography (AES–ICP) for most elements, with other instrumental techniques for elements such as carbon, oxygen and nitrogen.

Additional toughness tests were sometimes employed; in Part I, Schnadt impact tests and crack tip opening displacement (CTOD) tests were carried out on some initial series using non-standard specimens. Details are given in Ref. 8. More CTOD tests were made to examine strain ageing; they are reported in Chapter 12.

Most chapters are sub-divided as follows: an introductory section, which gives the chemical composition of the welds examined; the results of metallographic examination, divided into the structure of as-deposited un-reheated metal and coarse- and fine-grained reheated metal; mechanical properties (hardness, tensile properties and Charpy toughness); and usually finishing with a short discussion.

Addition and recovery of elements

Relatively simple formulations were used, typical of those used for basic, 25% iron powder electrodes but with no attempt to adjust the formulation to maintain weldability. Most elements were added to the dry mix as pure elements in powder form or as standard ferro-alloys, although in some cases, chemical compounds (e.g. ferrous sulphide for sulphur and ferrous phosphate for phosphorus) or minerals (rutile for titanium) were used. Nitrogen was added from nitrided ('Nitrel') manganese, containing 7%N. To achieve the higher purities required in the welds in Parts III–V, changes were made to some of the raw materials used; ferro-manganese was replaced by electrolytic manganese powder and rutile was replaced by titanium powder as a source of titanium, although where very small amounts were required in the presence of strong deoxidants, white titanium oxide was used. Chromium was normally added as the metal powder although, in practice, 70% ferrochrome is commonly employed. To vary oxygen below the usual value, magnesium was added to the dry mix, although recoveries of magnesium were close to the analytical detection limit. Magnesium also reduced sulphur and nitrogen contents.

When any element was added, care was taken to minimise its effects on recovery of other elements by suitably adjusting the formulation wherever possible. For instance, when nitrided manganese was used to add nitrogen, the proportion of electrolytic manganese was

Table 2.1 Recovery of added elements, best fit lines

Element	Method of addition	Recovery equation	R^2
*C (%)	Graphite	C = 0.042 + 0.124 (Graphite)	0.997
*Mn (%)	Fe–Mn	Mn = 0.055 + 0.195 (Fe–Mn)	0.998
*Mn (%)	Mn	Mn = 0.096 + 0.265 (Mn)	0.988
*Si (%)	Fe–Si	Si = 0.0151 + 0.0383 (Fe–Si) + 0.0072 (Fe–Si)^2	0.999
S (%)	FeS	S = 0.0068 + 0.05 (FeS)	0.997
P (%)	FePO$_4$	P = 0.0068 + 0.04 (FePO$_4$)	0.997
Mo (%)	Fe–Mo	Mo = 0.03 + 0.28 (Fe–Mo)	0.997
Cr (%)	Fe–Cr	Cr = 0.055 + 0.39 (Fe–Cr)	0.999
Ni (%)	Ni	Ni = 0.03 + 0.46 (Ni)	0.999
Cu (%)	Cu	Cu = 0.013 + 0.43 (Cu)	0.998
Zn (ppm)	Zn	Zn = 21.2 + 60.5 (Zn)	0.999
N (ppm)	Mn–N	N = 83 + 164.3 (Mn–N)	0.997
*Ti (ppm)	Rutile	Ti = 4.65 + 10.46 (rutile) – 0.32 (rutile)^2	0.996[†]
*Ti (ppm)	Ti	Ti = –1.68 + 95.3 (Ti) – 20.1 (Ti)^2 + 2.54 (Ti)^3	0.998
*Al (ppm)	Al	Al = –2.02 + 52.5 (Al) – 21.1 (Al)^2 – 3.7 (Al)^3	0.997[†]
*B (ppm)	Fe–B	B = –1.76 + 93.43 (Fe–B)	0.993
Nb (ppm)	Fe–Nb	Nb = 2.25 + 1150.5 (Fe–Nb)	0.999
V (ppm)	Fe–V	V = 3.2 + 2125 (Fe–V)	0.995

Notes: R is correlation coefficient.
Results are for a 1.4%Mn weld metal; they apply solely to the formulation and welding conditions used.
Values in brackets are additions to the coating, in wt%, to the dry mix.
* Balanced formulations.
† Other equations were developed giving higher (or lower) values of R^2.

reduced. It was not always possible, however, to achieve complete balancing; for example, additions of strong deoxidants such as titanium metal or magnesium reduced silicon out of the silicate binder employed after all the ferro-silicon had been removed from the formulation and this could give higher levels of silicon than were intended.

Recoveries of most elements examined into 1.4%Mn weld metal (except, naturally, manganese itself) are summarised in Table 2.1 and Fig. 2.3. It is strongly emphasised that these results are given for comparative purposes; they are valid only for the welding conditions employed and for the type of formulation used. The additions which were balanced are indicated in the Table.

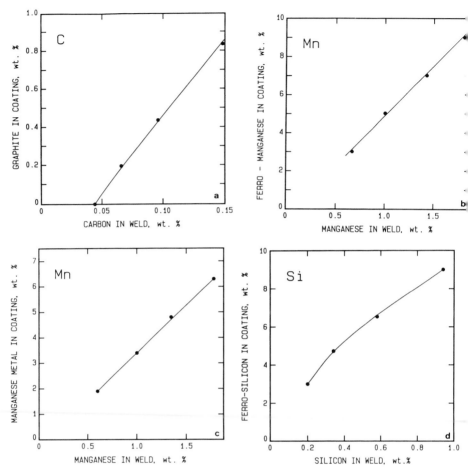

2.3 Recovery of elements into weld metal from additions to coating: *a*) Carbon from graphite; *b*) Manganese from 80% ferro-manganese; *c*) Manganese from electrolytic manganese; *d*) Silicon from 45% ferro-silicon; *e*) Sulphur from iron sulphide; *f*) Phosphorus from ferrous phosphate; *g*) Chromium from chromium powder; *h*) Molybdenum from 70% ferro-molybdenum; *i*) Nickel from nickel powder; *j*) Copper from copper powder; *k*) Zinc from zinc powder; *l*) Nitrogen from 7%N nitrided manganese; *m*) Titanium from rutile; *n*) Titanium from titanium powder; *o*) Aluminium from aluminium powder; *p*) Boron from 20% ferro-boron; *q*) Niobium from 65% ferro-niobium; *r*) Vanadium from 60% ferro-vanadium.

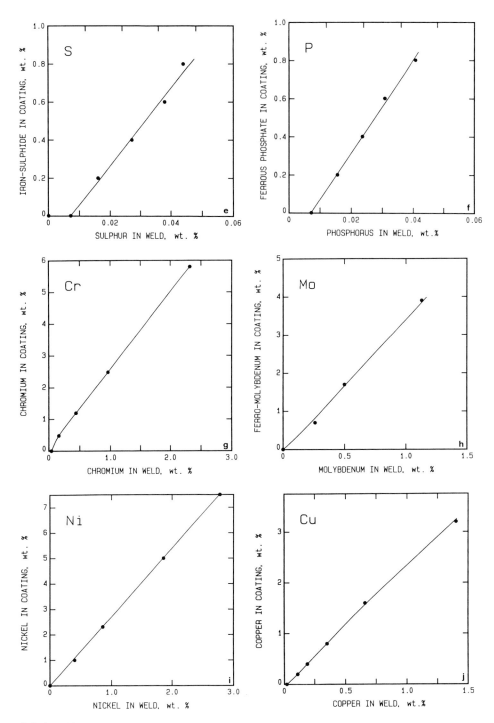

2.3 (cont.)

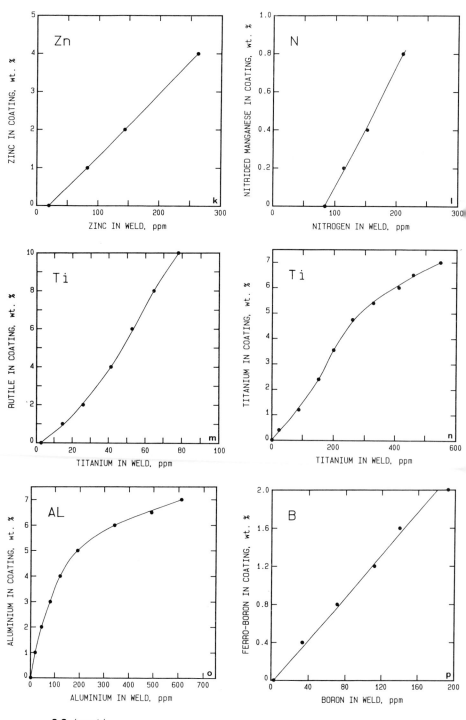

2.3 *(cont.)*

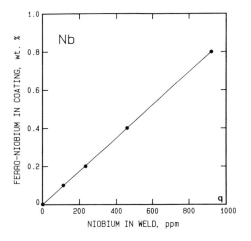
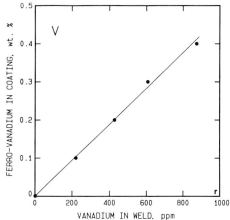

2.3 *(cont.)*

References

1 AWS A5.1-69: 'Specification for mild steel covered arc welding electrodes'. American Welding Society, 1969 [subsequently revised, 1981].

2 ISO 3690: 'Determination of hydrogen in deposited weld metal arising from the use of covered electrodes for welding mild and low alloy steels'. International Standards Organisation, 1977 [subsequently revised, 1983].

3 ISO 2560-1973: 'Covered electrodes for manual arc welding of mild and low alloy steels – code of symbols for identification'. ISO, 1973 [subsequently revised, 1974].

4 Davey T and Widgery DJ: 'A technique for the characterisation of weld metal microstructures'. IIW Doc II–A–389–76.

5 Anon: 'Guidelines for classification of ferritic steel weld metal microstructures in the light microscope'. *Welding in the World* 1986 (7/8) 24, 144–9; IIW DocII–A–664–85 (IX–1377–85).

6 Anon: 'Compendium of weld metal microstructures and properties'. Published by TWI for IIW, Cambridge, 1985.

7 Schnadt HM and Lienhard EW: 'Experimental investigation of the sharp notch behaviour of 60 steels at different temperatures and strain rates'. IIW Doc IX–343–63.

8 Evans GM: 'Effect of Mn on the microstructure and properties of all-weld-metal deposits'. *Welding J* 1980 **59**(3) 67s–76s; *Welding Res Abroad* 1983 **29**(1) 2–12; IIW Doc II–A–432–77; *Schweissmitteilungen* 1978 **36**(82) 4–19.

Part I

C–Mn weld metals

3

Manganese and carbon

Manganese and carbon are alloying elements deliberately added to mild steels and mild steel weld metals; both influence strength and toughness. However, in weld metals, excess carbon is undesirable because of its generally detrimental influence on weldability (Ref. 4, Ch. 1) – particularly cracking problems – so that manganese is the only practical way of controlling mechanical properties without recourse to more expensive alloying elements. Manganese also serves as a deoxidant, as does silicon (discussed in Chapter 6). However, silicon is also undesirable other than in small quantities (generally <0.5%), hence the particular importance of manganese.

The base composition studied in this and the following six chapters was one in which trace amounts of micro-alloying elements were present in sufficient quantities to allow formation of acicular ferrite should other conditions be suitable. The detailed effects of the various micro-alloying elements themselves are discussed in Parts IV and V.

Manganese

During the present, in some ways initial investigation, welds (Ref. 8, Ch. 2) were deposited from electrodes made using a coating factor of 1.70 (rather than the value of 1.68 later standardised) and using ferro-manganese to achieve nominal manganese contents of 0.6, 1.0, 1.4 and 1.8%. For easy reference, these compositions were coded A, B, C and D, respectively. The electrodes were made from the normal raw materials used for manufacture of good quality basic electrodes. The welds examined and discussed in this section differed from those deposited with what became the standard procedure in that an inter-pass temperature of 150 °C was used, rather than 200 °C selected as standard for all later welds, including those described in the second part of this chapter.

Table 3.1 Composition of weld metal having varying manganese content

Element, wt%	C	Mn	Si	S	P	N	O
Electrode code							
A	0.035	0.66	0.30	0.006	0.013	0.007	0.049
B	0.038	1.00	0.30	0.005	0.014	0.010	0.046
C	0.049	1.42	0.34	0.005	0.013	0.009	0.041
D	0.051	1.82	0.34	0.006	0.017	0.009	0.039

Note: typical impurity levels were 0.03%Cr, Ni, Cu, 0.005%Mo, 120 ppm V, 55 ppm Ti, 20 ppm Nb, 5 ppm Al and 2 ppm B.

The chemical compositions of the weld metals with varying manganese contents are detailed in Table 3.1. Analysis of welds tested in the stress-relieved condition gave similar results and typical impurity levels are given as a footnote to the table.

In general, the compositions were well balanced and invariant elements showed reasonably constant levels, except that carbon, silicon and phosphorus increased slightly as manganese was increased, whilst oxygen decreased. The changes in oxygen (and also carbon) are consistent with manganese acting as a deoxidant at this level of addition. Because of this relatively constant behaviour, analyses given for some later series are less detailed and reference should be made to original papers for complete results. Similarly, if there were no significant effects to report, some results on other features and properties are less detailed in subsequent chapters.

Metallographic examination

A transverse macrosection of a typical weld deposit is shown in Fig. 3.1, which should be compared with the idealised sketch in Fig. 2.1. Measurements of the proportions of columnar, coarse and fine grained regions were made on such cross sections; these are summarised graphically in Fig. 3.2. The widths of the remaining columnar zones in each run varied irregularly, as did the widths of the coarse grained regions, so that no influence of manganese was apparent. In some runs the columnar region was completely absent. In the regions where notches of Charpy specimens would be located, the measured proportions of the different zones are given in Table 3.2. Typical microstructures of the three zones of deposit of all four manganese levels are shown in Fig. 3.3–3.5.

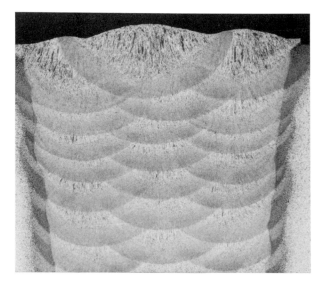

3.1 Typical cross section of weld, multirun deposit.

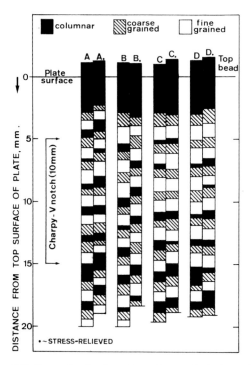

3.2 Zone distribution along the vertical plane for duplicate welds of varying manganese content.

Table 3.2 Percentage of different zones at the location of the Charpy V notch for welds of varying manganese content

Weld identity	A, 0.66Mn		B, 1.0Mn		C, 1.4Mn		D, 1.8Mn		Mean
Zone	AW	SR	AW	SR	AW	SR	AW	SR	
Columnar	18,	32	23,	19	22,	12	11,	20	20
Coarse grained	35,	24	34,	35	34,	37	34,	37	34
Fine grained	47,	42	43,	46	44,	51	55,	45	46

Note: AW – as-welded; SR – stress relieved at 580 °C.

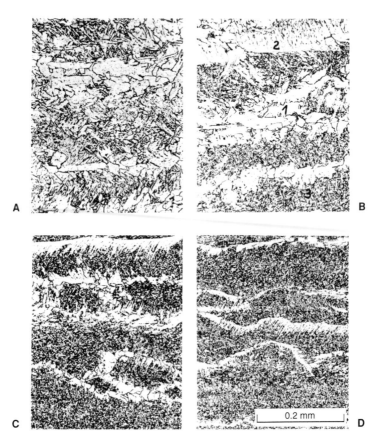

3.3 Microstructure of top columnar beads of welds A–D (in weld B, 1 – grain boundary ferrite, 2 – ferrite with aligned second phase, 3 – acicular ferrite).

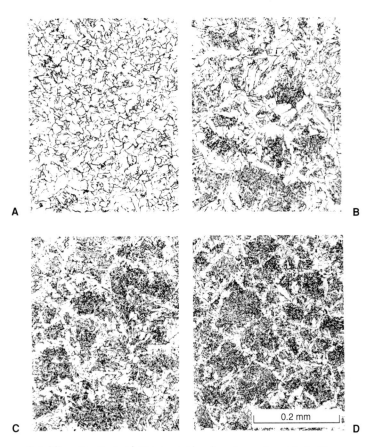

3.4 Microstructure of coarse grained regions of welds A–D.

Columnar region

The proportions of the three constituents, grain boundary (pro-eutectoid) ferrite, ferrite with aligned second phase (side plates) and acicular ferrite, were obtained by point counting 500 fields by two investigators at a magnification of ×200. The results, summarised in Fig. 3.6, show a clear trend for both grain boundary ferrite and (to a lesser degree) for ferrite with aligned second phase to be replaced by acicular ferrite as the manganese content of the weld was increased. Stress relief at 580 °C had no effect on the proportions of the con-stituents revealed by this examination. Replicas of top beads exam-ined by transmission electron microscopy showed a reduction in

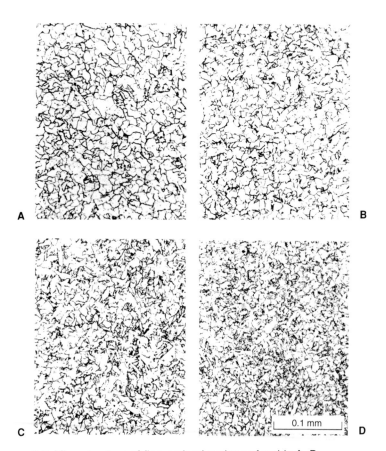

3.5 Microstructure of fine grained regions of welds A–D.

linear intercept ferrite grain size as the manganese content was increased, see Table 3.3. At the highest manganese level, the grain size approximated to that of the acicular ferrite constituent. This examination also showed that there was a much less sharp distinction between ferrite with aligned second phase and acicular ferrite than is apparent in the light microscope.

In all replicas, small and widely dispersed areas of retained austenite were seen. Only in weld D of the highest manganese content was there sufficient (1%) to allow its detection and estimation by X-ray diffraction. Stress relieving caused this austenite to transform to ferrite and grain boundary carbides.

Any martensite which may have formed within the retained austenite was difficult to detect because of the very fine scale of the

Table 3.3 Mean linear intercept ferrite grain size of top beads of varying manganese content at ×2500 magnification

Weld identity	Mean intercept ferrite grain size, μm	
	As-welded	Stress-relieved, 580 °C
A, 0.66 Mn	3.3	4.0
B, 1.0 Mn	2.9	2.6
C, 1.4 Mn	1.7	1.7
D, 1.8 Mn	1.0	1.6

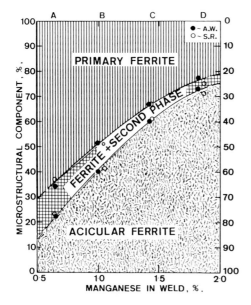

3.6 Influence of manganese on microstructural constituents of top, unreheated weld beads.

structure; it was certainly much finer than that found in submerged-arc welds.[1] Furthermore, it was not possible to identify such areas as either lath or twinned martensite or to quantify them.

Coarse-grained region

In the coarse-grained high temperature region, increasing manganese content again increased the proportion of acicular ferrite, so that

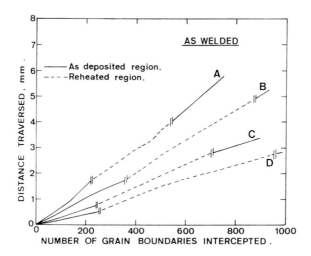

3.7 Grain boundaries intercepted on traversing as-deposited and reheated regions.

pro-eutectoid ferrite more obviously delineated prior austenite boundaries (Fig. 3.4) and hence accentuated the coarse grained nature of this zone. With the lowest manganese level (weld A), it was difficult to identify fusion boundaries, although segregation bands could be identified near such boundaries on altering the focus of the microscope.

Traverses across as-deposited and reheated regions made in the SEM to intercept ferrite boundaries showed a steady decrease in grain size as the manganese content was increased (Fig. 3.7). It was also apparent that there was a change in slope of the traverse lines when boundaries were crossed. It was not, however, possible to detect boundaries between the intercritically and fully reheated regions in this way.

Fine-grained region

The fine grained low temperature regions of all welds (Fig. 3.5) were essentially equiaxed (Table 3.4). Plotting the reciprocal of the square root of the mean intercept grain size against manganese content (Fig. 3.8) showed a linear relationship. These results are almost identical with those obtained by Tuliani[2] for the reheated runs of submerged-arc welds deposited at a higher heat input.

Table 3.4 Linear intercept results from fine grained zones of welds of varying manganese content

Weld identity	Intercepts per mm	Ratio	Mean intercept, mm^{-1}	Mean interval, l, μm	l$^{-1/2}$, mm$^{-1/2}$
A, 0.66 Mn	H 155 / V 146	1.06	150	6.7	12.2
B, 1.0 Mn	H 173 / V 171	1.01	172	5.8	13.1
C, 1.4 Mn	H 199 / V 208	0.96	203	4.9	14.3
D, 1.8 Mn	H 237 / V 253	0.94	245	4.1	15.6

Note: H – horizontal traverse; V – vertical traverse.

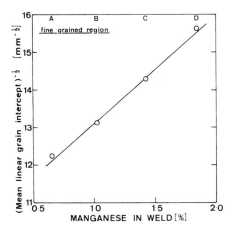

3.8 Effect of manganese on the grain size of fine-grained regions.

Mechanical properties

Hardness tests were not made in this series, although effects of manganese on the hardness of as-deposited weld metal can be assessed from the values in Tables 4.2 and 4.11 of the next chapter.

Tensile tests

The results of tensile tests of specimens in both the as-welded and stress-relieved conditions are given in Table 3.5. Both strength properties increased linearly with manganese content, as shown in Fig. 3.9; ductility values fell slightly. Strength values were slightly reduced and elongation values slightly increased by stress relief heat treatment.

Best fit straight lines, appropriate solely to the welding conditions used, gave the following relationships:

$$YS_{aw} = 314 + 108 \ Mn \qquad [3.1]$$

$$TS_{aw} = 394 + 108 \ Mn \qquad [3.2]$$

$$YS_{sr} = 311 + 89 \ Mn \qquad [3.3]$$

$$TS_{sr} = 390 + 98 \ Mn \qquad [3.4]$$

From these it can be seen that in the as-welded condition 0.1% manganese increased both yield and tensile strength by approximately $10 \ N/mm^2$ and that stress relieving reduced strength by amounts which increased as the manganese content increased.

Toughness tests

In the as-welded condition, three types of toughness test were carried out: Charpy V notch, CTOD and Schnadt impact test. Only details of the Charpy tests are recorded here, as the CTOD notches were not

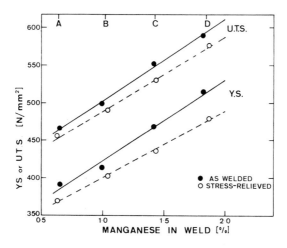

3.9 Effect of manganese on the strength properties of multirun deposits.

Table 3.5 Tensile test results for welds of varying manganese content

Condition	Electrode	Yield strength, N/mm²	Tensile strength, N/mm²	Elongation, %	Reduction of area, %
AW	A, 0.66 Mn	390	470	32	81
	B, 1.0 Mn	410	500	31	81
	C, 1.4 Mn	470	550	29	79
	D, 1.8 Mn	510	590	28	77
SR	A, 0.66 Mn	370	460	35	81
	B, 1.0 Mn	400	490	31	81
	C, 1.4 Mn	440	530	32	79
	D, 1.8 Mn	480	580	27	77

Note: AW – as-welded; SR – stress relieved.

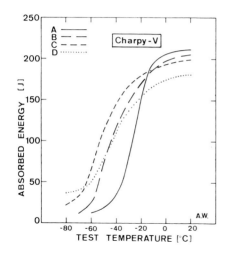

3.10 Charpy V transition curves for welds of varying manganese content, as-welded.

(as is now customary) extended by fatigue cracking and the Schnadt test is no longer in general use; these results can be found in Ref. 8 of Chapter 2. The Charpy tests were carried out on specimens as-welded, after stress-relief (2 hours at 580 °C) and in the strain aged condition (compressed 10% and aged 0.5 hours at 250 °C – see Chapter 7 for further details).

In Fig. 3.10 the transition curves for as-welded Charpy tests are reproduced; Fig. 3.11 shows the Charpy energy values plotted against

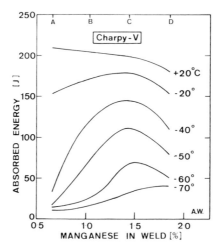

3.11 Effect of manganese on as-welded Charpy V transition temperature.

manganese content for the stated temperatures. These graphs reveal clearly that toughness, in terms of Charpy transition temperature, exhibited an optimum level (i.e. a minimum transition temperature) with a manganese content close to 1.4%. However, at very low temperatures, below the toughness transition, increasing manganese appeared to increase lower shelf Charpy toughness, whilst Charpy upper shelf values (those at 20 °C) were reduced by increasing the manganese level. The benefit of manganese up to the optimum value of 1.4% was in lowering the transition temperature between fully ductile and fully brittle fracture. Although two Charpy transition temperatures were determined during the present work, subsequent discussion concentrates on the 100 J temperature. However, the 28 J transition and the individual Charpy curves can be found in the original papers.

The corresponding Charpy data for the stress-relieved condition, see Fig. 3.12 and 3.13, show similar results with relatively small changes in transition temperatures from the as-welded values. For example, the change in the temperature for 100 J was less than 10 °C, stress relief being marginally beneficial at low manganese levels and marginally harmful at high levels.

Strain ageing, on the other hand, see Fig. 3.14 and 3.15, produced a marked worsening of toughness and altered the influence of manganese. The increase in the 100 J temperature due to strain ageing averaged about 35 °C, whilst manganese continued to improve tough-

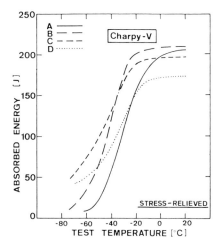

3.12 Charpy V transition curves for welds of varying manganese content, stress-relieved.

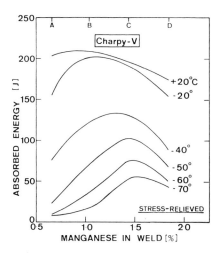

3.13 Effect of manganese on stress-relieved Charpy V transition temperature.

ness up to the highest level of 1.8%. This was because the greatest shift in transition temperature of 41 °C was found with the hitherto optimum 1.4%Mn weld metal and the lowest (24 °C) occurred with 1.8%Mn. These results are discussed further in Chapter 7.

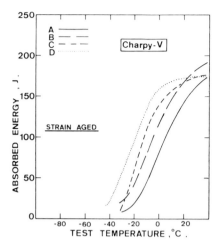

3.14 Charpy V transition curves for welds of varying manganese content, strain-aged.

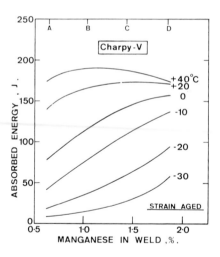

3.15 Effect of manganese on strain aged Charpy V transition temperature.

Carbon and manganese

In this second series, repeated at different carbon levels, welds were deposited using the standard interpass temperature of 200 °C, the carbon content being varied from 0.04 to 0.15% in four series of

Table 3.6 Composition of weld metal having varying C content at different Mn level

Element, wt%		C	Mn	Si
Mean C level, %	Electrode code			
0.044	A	0.045	0.65	0.30
	B	0.044	0.98	0.32
	C	0.044	1.32	0.32
	D	0.045	1.72	0.30
0.064	A	0.059	0.60	0.33
	B	0.063	1.00	0.35
	C	0.066	1.35	0.37
	D	0.070	1.77	0.33
0.096	A	0.099	0.65	0.35
	B	0.098	1.05	0.32
	C	0.096	1.29	0.30
	D	0.093	1.65	0.33
0.147	A	0.147	0.63	0.40
	B	0.152	1.00	0.41
	C	0.148	1.40	0.38
	D	0.141	1.76	0.36

Note: all deposits contained 0.005–0.008%S and 0.007–0.009%P and typical impurity levels of 0.03%Cr, Ni, Cu, 0.005%Mo, 120 ppm V, 55 ppm Ti, 20 ppm Nb, 5 ppm Al and 2 ppm B.

welds,[3] each series having the same four manganese levels used in the previous section and electrodes giving the possibility for acicular ferrite formation. Welds were tested in the as-welded condition (tensile specimens having been given a hydrogen-removal treatment at 250 °C); tests on these welds after PWHT are discussed in Chapter 5. The compositions of the deposits given in Table 3.6 show that the carbon content of any weld was no more than 0.006%C from the mean value and the spread of manganese contents no more than 8% of the mean value for each series. The contents of S and P were uniform, but a slight increase in silicon (no more than 0.1%) was encountered as the carbon content was increased.

Metallography

Conventional metallography on specimens from the test welds is described below. Electron microscopy was carried out on selected deposits for comparison with welds in the PWHT condition; these

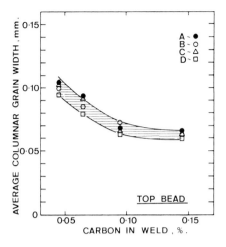

3.16 Influence of carbon on mean width of columnar grains in the as-deposited microstructure.

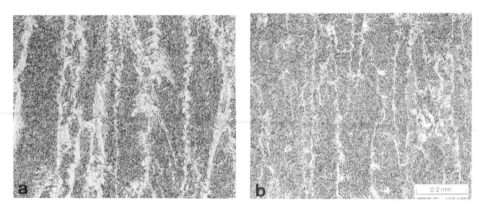

3.17 As-deposited microstructure of 1.4%Mn welds with: *a*) 0.045%C; *b*) 0.148%C.

tests are described in Chapters 5 and 17. The main effect of carbon was to alter the microstructure seen at high magnification, although in all deposits carbon decreased the width of the columnar grains (Fig. 3.16) and gave thinner ferrite veins at the prior austenite boundaries (Fig. 3.17).

The results of point counting the final, as-deposited beads at ×500 magnification (as recommended by the IIW method, Ref. 5, Ch. 2)

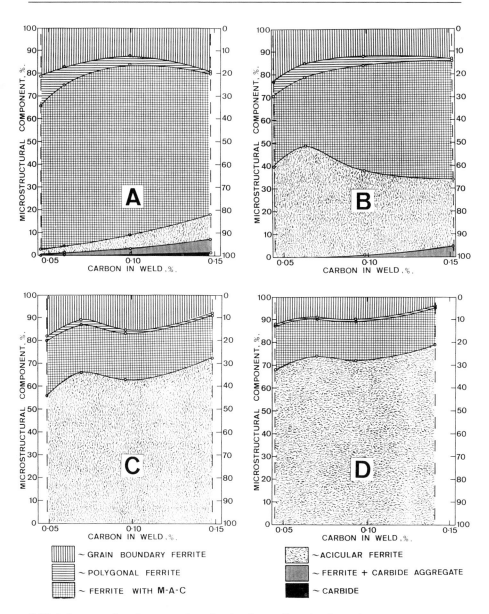

3.18 Influence of carbon on microstructural constituents at varying manganese levels.

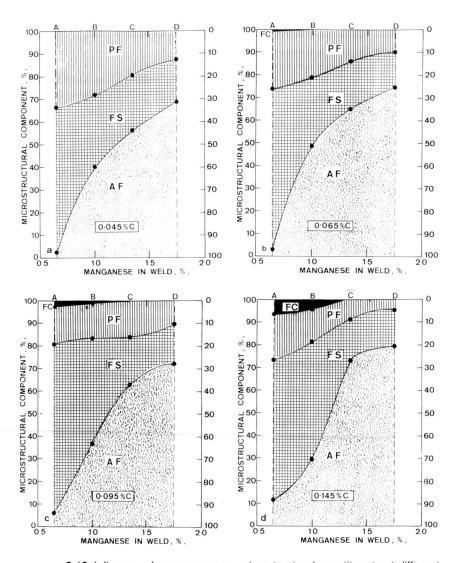

3.19 Influence of manganese on microstructural constituents at different carbon levels.

are shown in terms of carbon content at different manganese levels in Fig. 3.18 and in terms of manganese at different carbon levels in Fig. 3.19. Increasing carbon reduced the proportions of both grain boundary and polygonal ferrite in the as-deposited microstructure. At the two lowest carbon levels, the grain boundary ferrite was replaced by ferrite with aligned second phase, ferrite/carbide aggregate and, at

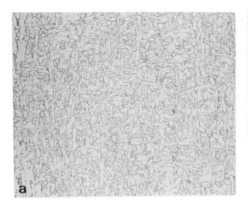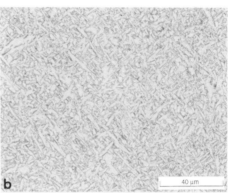

3.20 Acicular ferrite structure in as-deposited weld metal with 1.4%Mn and: *a*) 0.045%C; *b*) 0.145%C.

the very lowest level, by carbides and acicular ferrite. At higher carbon levels ferrite/carbide aggregate and carbides were absent and acicular ferrite increased as carbon was increased, although after the first addition of carbon, acicular ferrite again was reduced at the 1%Mn level.

Re-plotted against manganese, Fig. 3.19 (in which grain boundary and polygonal ferrite are combined as primary ferrite) shows that manganese reduced all constituents except acicular ferrite. However, the rate of increase in acicular ferrite with manganese addition fell as manganese content was increased, and for the two highest carbon contents there was little difference in acicular ferrite content between the nominal 1.4% and 1.8%Mn welds.

Under higher magnification in the SEM, increasing carbon was found to increase substantially the proportion of retained phases – martensite and retained austenite – between the laths of acicular ferrite. Increasing carbon also increased the aspect ratio of the ferrite laths (Fig. 3.20), giving a classical Widmanstätten structure at highest carbon level.

Changes in the coarse-grained reheated microstructures with carbon mirrored those in the as-deposited microstructure. Increasing carbon reduced the width of grain boundary ferrite and increased the amount of acicular ferrite – particularly at the higher manganese levels. Carbon significantly refined the ferrite grain size within the fine-grained region, see Fig. 3.21 and 3.22, an effect similar to that of manganese. In addition, carbon gave a more duplex appearance to the microstructure (Fig. 3.21(*b*)), and phases tended to separate from the ferrite, particularly in segregate bands. Deep etching in a bromine/methanol mixture and SEM examination (by BSC) revealed

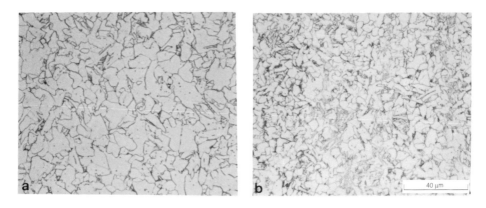

3.21 Microstructure of fine grained regions of deposits with: *a)* 0.045%C; *b)* 0.145%C.

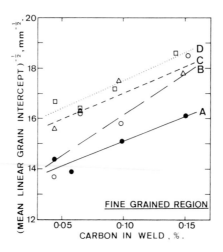

3.22 Influence of carbon on grain size within the refined region of welds of varying manganese levels.

several constituents, all of which increased as the carbon content was increased, as shown in Fig. 3.23.

Mechanical properties

Hardness

As with manganese, carbon increased the hardness of the top bead in a linear manner (Fig. 3.24). Hardness traverses along weld centre-

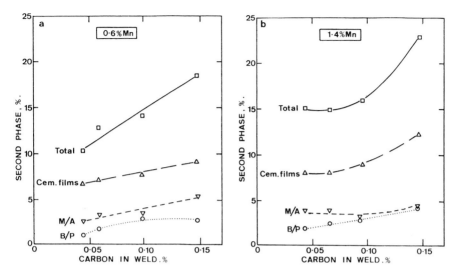

3.23 Influence of carbon on constituents within the refined region of welds of: *a*) 0.6%Mn; *b*) 1.4%Mn (Cem – cementite, M/A – martensite/austenite, B/P – bainite/pearlite).

lines, typified by Fig. 3.25, gave a wave-like pattern, with the top bead ~30 HV5 harder than the general level of the underlying weld metal.

Tensile tests

Tensile properties in the as-welded condition (i.e. after 14 hours at 250 °C) are given in Table 3.7. Figure 3.26 shows that strength properties were proportional to the carbon content as well as to manganese (Fig. 3.9) and the following two equations were developed:

$$YS_{aw} = 335 + 439\ C + 60\ Mn + 361\ (C.Mn) \qquad [3.5]$$

$$TS_{aw} = 379 + 754\ C + 63\ Mn + 337\ (C.Mn) \qquad [3.6]$$

Compared with the equations initially developed for manganese alone (Eq. [3.1] and [3.2]), the more complex equations show a smaller factor for the influence of manganese (about 60 compared with about 100). Some of this is a result of the tendency for the carbon content of the earlier welds to increase as the manganese content was increased (Table 3.1) and some is accounted for by the positive *C.Mn* term.

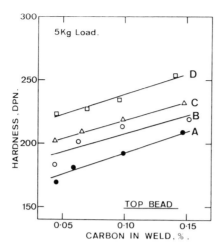

3.24 Influence of carbon on as-welded top bead hardness of deposits of varying manganese levels.

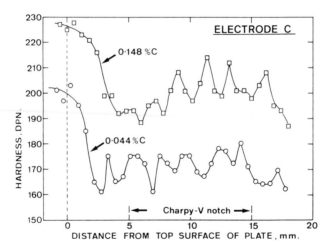

3.25 Hardness variation along the centreline of welds with 1.4%Mn and 0.045 or 0.145%C.

Ratios of yield to tensile strength varied between 0.84 and 0.90 with no consistent pattern of variation. Ductility values fell only slightly with carbon within the range of C and Mn contents examined.

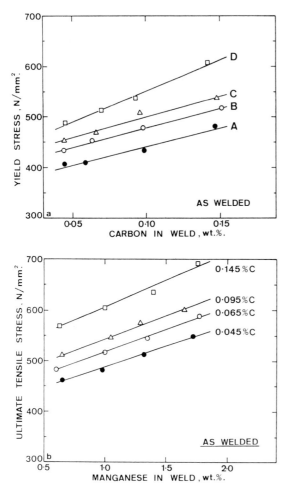

3.26 Effect of carbon on as-welded strength properties of multirun deposits with varying manganese content: *a*) Yield strength; *b*) Tensile strength.

Charpy tests

Transition curves are shown in Fig. 3.27. Although carbon appeared to increase energy levels at the lower shelf and decrease them on the upper shelf, it had little effect on transition temperatures, as exemplified by Fig. 3.28. Carbon was mildly beneficial at the lowest manganese content, mildly harmful at the highest and gave an optimum level of about 0.09% at intermediate levels. Increasing the carbon

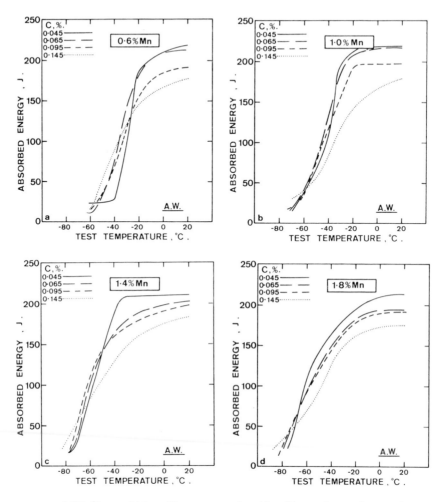

3.27 Charpy V transition curves of welds with varying carbon and manganese content, as-welded.

content reduced the slope of the Charpy transition curve and hence decreased the scatter of values within the transition region; this is shown for 1.4%Mn deposits in Fig. 3.29. With the lowest carbon content, the transition was so steep that bimodal fracture behaviour (i.e. fracture being either completely ductile or completely brittle within the transition range[4]) occurred at −40 °C.

In Fig. 3.30, plots of Charpy energy against manganese for different carbon levels show maximum energy values close to 1.4%Mn,

Table 3.7 As-welded tensile test results on welds of varying carbon and manganese content

C level, %	Electrode code	Yield strength, N/mm²	Tensile strength, N/mm²	Elongation, %	Reduction of area, %
0.044	A	410	460	35	79
	B	430	480	36	79
	C	450	510	32	79
	D	490	550	30	76
0.064	A	410	480	31	81
	B	450	520	32	81
	C	470	540	29	79
	D	510	590	28	78
0.096	A	430	510	32	79
	B	480	550	30	79
	C	510	580	31	78
	D	540	610	28	74
0.147	A	480	570	33	76
	B	520	600	27	75
	C	540	640	27	76
	D	610	690	26	72

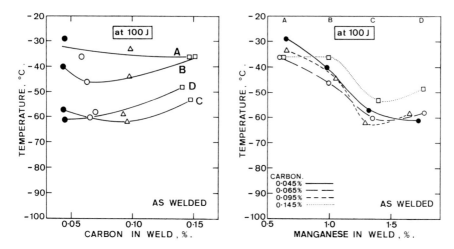

3.28 Influence of carbon and manganese on 100 J Charpy transition temperature.

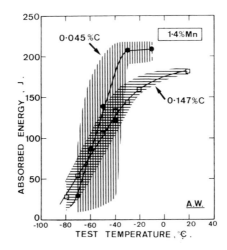

3.29 Charpy transition curves for 1.4%Mn welds with low and high carbon content showing steeper transition curve for the lower carbon deposit.

except for the lowest carbon series, where it was 1.7%Mn. Carbon, therefore had little effect on the optimum manganese level of 1.4%.

Discussion

The influence of manganese on hardness and strength of low carbon steels and weld metals is well documented.[3,5-7] The numerical influence on tensile strength – 0.1%Mn increasing strength by ~10 N/mm^2 – is similar to values obtained by other workers[7,8] on submerged-arc and CO_2 weld metals. Although carbon has a greater influence on strength than manganese on a percentage basis, the useful range within which it can be varied limits its possible usefulness. The relatively small detrimental influences of manganese and carbon on ductility are in line with their effects in low carbon steels.[5]

In a weld metal favourable to formation of acicular ferrite, addition of manganese replaced relatively coarse primary ferrite in as-deposited and coarse-grained reheated zones by much finer acicular ferrite instead of by (coarse) ferrite with aligned second phase. This grain refinement adds to the influence of manganese on increasing strength and hardness and, more importantly, improves toughness where (in the absence of such refinement) a deterioration would have been expected.

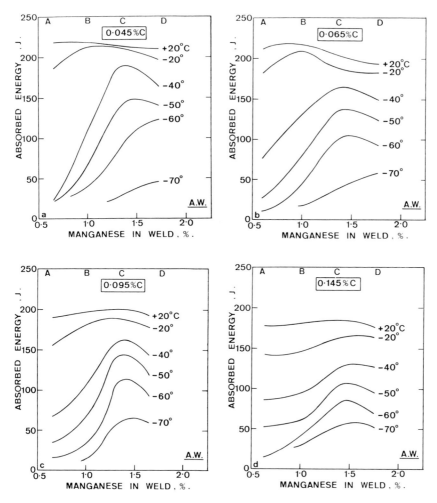

3.30 Effect of manganese on as-welded Charpy V transition temperature at varying carbon levels.

As can be seen from Fig. 3.6 and 3.19, the rate of increase in the content of acicular ferrite with manganese was not linear but fell as the amount of manganese increased, particularly above 1.4%. This non-linearity, coupled with the increased amount of microphases, was probably responsible for the strengthening influence of manganese above 1.4% outweighing its refining effect and so reducing toughness. Thus, the optimum manganese content was 1.4%, despite continuing microstructural refinement at the higher level. The results are in line with the observation of Dolby[9] that, in weld metals capable of developing acicular ferrite, as-welded toughness improves and then deteriorates as the manganese content increases.

A comparison of the earlier (Fig. 3.6) quantitative microstructural results with later results in Fig. 3.19 (with carbon at the lowest level) shows some apparent discrepancies. Some of these can largely be resolved by taking account of the slight but systematic increase in carbon with manganese in the earlier series, use of a lower interpass temperature of 150 °C in the earlier series compared with 200 °C in the later series, and point counting the earlier results with a magnification of only ×200.

Over the range studied, the influence of carbon on toughness was less than that of manganese, although it gave an optimum level in the range 0.07–0.09% for the optimum 1.4%Mn electrode. At low manganese contents, carbon increased the proportions of constituents other than acicular ferrite, whilst at about 1.4%Mn the first increase from 0.04 to 0.06%C gave a useful increase in acicular ferrite. Other work[10] has shown that the toughness properties of specially manufactured E7018 electrodes (i.e. similar to those examined here, but with ultra low carbon contents) gave poor Charpy toughness values at −20 °C, even though the upper shelf values at ambient temperature were extremely good.

Svensson and Grethof[11] employed a slightly different range of compositions (0.03–0.12%C, 0.8–2%Mn, with about 120 ppm Ti) from that examined in the present work. They found a deterioration in toughness with excessive alloying, which was associated with segregation of microphases (particularly retained austenite) into almost continuous bands. They also found indifferent toughness at the lowest level of 0.03%C, which they ascribed to higher nitrogen content.

The reasons for the non-linear increase in the proportion of acicular ferrite with manganese and the maximum in toughness at about 1.4%Mn were not apparent from the metallographic studies carried out at this stage of the investigation. The only obvious feature was a very small increase in the amount of residual austenite. This phase will tie up manganese and carbon otherwise available to form more acicular ferrite. Also, although residual austenite is not, by itself, a reason for poor toughness, it will have been transformed to some form of carbide or martensite in the low temperature reheated regions of the underlying, reheated runs, which were the runs tested. Retained austenite may also transform to martensite during cooling to the low sub-zero temperatures required for some types of testing (including impact testing) and hence be harmful in its own right.

The optimum manganese and carbon contents with regard to toughness are related to transition temperature. The Charpy upper shelf toughness appears to worsen continuously as both manganese and carbon are increased from the lowest to the highest levels. This is understandable, in that upper shelf levels of weld metals of similar inclusion content (demonstrated by the similar oxygen and sulphur contents of all deposits), will be controlled principally by strength, which is progressively increased by both elements.

At the lower shelf, Charpy toughness improved slightly as both manganese and carbon were increased. This is probably a reflection of Charpy propagation energy increasing with the strength of the weld metal and refinement of microstructure outweighing any detrimental influence of carbides.

Although carbon is a more potent strengthening element than manganese, its potential for variation in a practical MMA weld is much more limited because of its generally adverse effect on cracking problems (Ref. 4, Ch. 1).

The results of tests on strain aged weld metal show a moderate loss of toughness; there is further comment on this in Chapter 11.

References

1 Garland JG and Kirkwood PR: 'The notch toughness of submerged arc weld metal in micro-alloyed structural steels'. IIW Doc IX–248–63.
2 Tuliani SS: 'The role of manganese in mild steel submerged arc weld metal'. CEGB, Marchwood England, Sept 1972.
3 Evans GM: 'Effect of carbon on the microstructure and properties of C–Mn all-weld-metal deposits'. *Welding J* 1983 **62**(11) 313s–20s; *Welding Res Abroad* 1983 **29**(1) 58–70; IIW Doc II–983–82; *Schweissmitteilungen* 1982 **40**(99) 17–31.
4 Bailey N: 'Bimodality revisited – split behaviour of weld metal'. *TWI Res Bull* 1991 **32** 110–15.
5 Carpenter H and Robertson JM: 'Metals'. OUP, London, 1939, pp 1081–85.
6 Rees WP, Hopkins BE and Tipler HR: 'Tensile and impact properties of iron and some iron alloys of high purity'. *JISI* 1951 **169** 157–68.
7 Brain AG and Smith AA: 'Mechanical properties of CO_2 weld metal'. *Brit Weld J* 1962 **9** 669–77.
8 Bailey N and Pargeter RJ: 'The influence of flux type on the strength and toughness of submerged arc weld metal'. TWI Report Series, 1988.
9 Dolby RE: 'Factors controlling weld toughness – the present position. Part 2, Weld metals'. TWI Res Report, 14/1976/M.

10 Sagan SS and Campbell HC: 'Factors which affect low-alloy weld metal notch toughness'. *Weld Res Council Bull* No 58, April 1960.

11 Svensson L-E and Grethof B: 'Microstructure and impact toughness of C–Mn weld metals'. *Welding J* 1990 **69**(12) 454s–61s.

4

Effects of welding process variables

In manual welding, it is inevitable that welding conditions vary slightly within each run and between runs and that they differ for different types of weld. In this chapter, effects of changing the welding parameters within (and even beyond) their normal limits are examined for the C–Mn weld metals described in the previous chapter. Also, the results of tests are reported which were made to compare the two major types of testing available world-wide – the ISO 2560 method (Ref. 3, Ch. 2) used throughout this project and the AWS A 5.1–69 (Ref. 1, Ch. 2). Use of the latter is widespread throughout the American continent and, for some applications, it was used throughout the world at the time the work was done. Both these specifications have been modified since the tests were carried out, as will be explained in the appropriate section.

Interpass temperature

The interpass temperature – the temperature immediately before each weld run after the first is deposited – is more important in multipass welding than the preheat temperature (the temperature immediately before the first run is deposited). In ISO 2560:1973 (Ref. 3, Ch. 2), the standard for qualification of electrodes, the only stipulation on interpass temperature was a maximum value of 250 °C, well suited to welding the relatively short block (Fig. 2.2) used to provide sufficient weld metal for the necessary tests. In structural welding of large components, much lower values can be encountered, particularly when preheat is not used.

Electrodes of four different manganese levels (coded A, B, C and D, as in Chapter 3) were used;[1] interpass temperatures were varied between 20 and 300 °C for each electrode. Tests were carried out in the as-welded condition; toughness was assessed by Charpy tests only. The analysis results given in Table 4.1 show similar compositions to

Table 4.1 Influence of interpass temperature on composition of weld metal having varying Mn content

Interpass temperature, °C	Electrode code	Element, wt%		
		C	Mn	Si
20	A	0.036	0.60	0.31
	B	0.043	1.03	0.30
	C	0.046	1.39	0.31
	D	0.050	1.79	0.31
150	A	0.039	0.58	0.28
	B	0.047	1.00	0.31
	C	0.046	1.40	0.31
	D	0.049	1.77	0.31
240	A	0.037	0.60	0.28
	B	0.042	1.02	0.32
	C	0.047	1.37	0.31
	D	0.047	1.75	0.32
300	A	0.042	0.55	0.25
	B	0.045	0.99	0.30
	C	0.047	1.32	0.27
	D	0.047	1.66	0.28

Note: all welds contained 0.007–0.008%S and 0.010–0.015%P; typical levels of other elements are given in Table 3.1.

Table 4.2 Influence of interpass temperature on cooling time and hardness of top run beads having varying manganese content

Interpass Temperature, °C	20		150	240	300
Cooling time, $T_{8/5}$, sec	4		6	8	10
Electrode code	Hardness, HV5				
A	197		192	168	155
B	218		195	181	184
C	243		205	200	197
D	250		243	215	208

the welds made with the same electrodes in Chapter 3 (Table 3.1). There was little variation in composition as a result of differences in interpass temperature, except that the content of the deoxidants manganese and silicon fell slightly with increasing interpass temperature, particularly at the highest temperature. Further tests on alloyed weld metals, intended to give yield strengths above 690 N/mm², are described in Chapter 16.

Thermocouples were harpooned into the final runs of selected weld deposits to determine how the interpass temperature influenced weld cooling, assessed by the 800–500 °C cooling time ($T_{8/5}$). These results (obtained at the Institut de Soudure) are included in Table 4.2.

Metallography

Metallographic examination showed a marked difference in macrostructure as the interpass temperature was increased. Figure 4.1 shows how increasing temperature considerably reduced the proportion of columnar structure and increased the amount of refined structure, particularly in the central beads of the deposit.

Measurements of the top weld beads and their HAZs on the underlying runs showed scarcely any change in size of the cross sectional area of the beads themselves but a marked increase in the proportion of the supercritically reheated (recrystallised) area, particularly with the highest interpass temperature (Fig. 4.2).

Increasing the interpass temperature increased the proportion of both fine and coarse grained regions at the expense of columnar (Fig. 4.3). With the highest interpass temperature of 300 °C, very little columnar structure remained. This finding is in agreement with investigations into the possibility of refining parent steel HAZs to improve toughness for as-welded repairs[2] or to reduce the risk of reheat cracking during PWHT after welding Cr–Mo–V steels.[3] These investigations have shown that preheating and maintaining relatively high interpass temperatures are helpful in achieving a high degree of HAZ refinement.

The virtual elimination of the columnar microstructure also succeeded in giving a more uniform hardness in the weld metal. This is shown in the hardness traverses down weld centrelines, reproduced in Fig. 4.4. Hardness itself was reduced by increasing the interpass temperature, see Table 4.2, which gives the top run hardnesses and the corresponding measured 800–500 °C cooling times. Mean top run hardness values decreased by an average of about 40 HV5 as the interpass temperature was increased over the full range studied.

Quantitative metallography showed that the proportions of acicular ferrite in the as-deposited microstructure with all electrodes decreased as the interpass temperature was increased, whilst the amounts of grain boundary ferrite and, to a smaller degree, of ferrite with aligned second phase increased (Fig. 4.5). Micrographs of the

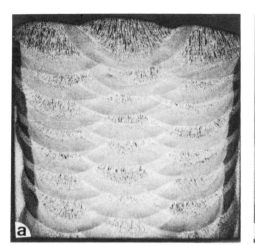 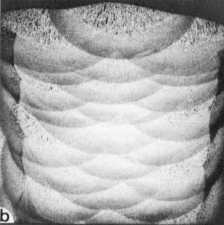

4.1 Macrostructure of welds deposited with: a) 20 °C; b) 300 °C interpass temperature.

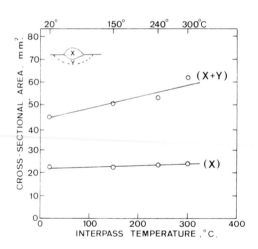

4.2 Influence of interpass temperature on weld bead (X) and recrystallised HAZ (Y) size.

1.5%Mn deposit (Fig. 4.6), illustrate the influence of interpass temperature on the proportions of the constituents and also show that a high interpass temperature coarsened the acicular ferrite lath size.

The microstructures of the coarse-grained reheated regions (Fig. 4.7), were also coarsened by increasing the interpass temperature.

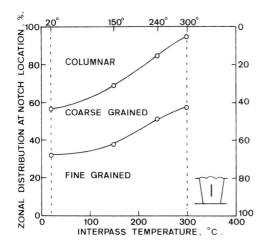

4.3 Influence of interpass temperature on zonal distribution at the Charpy notch position.

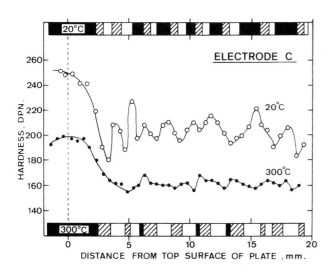

4.4 Hardness traverses down the centre of welds made with electrode C (1.4%Mn) deposited with 20 or 300 °C interpass temperature.

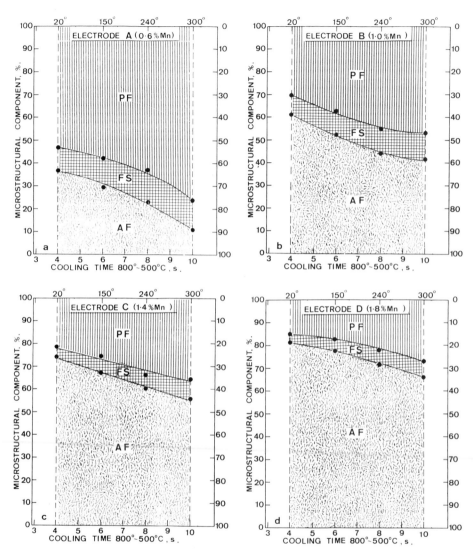

4.5 Effects of interpass temperature and cooling time on microstructure of welds containing: a) 0.6%Mn; b) 1.0%Mn; c) 1.4%Mn; d) 1.8%Mn.

This coarsening applied both to the ferrite outlining the prior austenite grains and to the finer ferrite within them. It was also noted that the size of the coarse areas appeared to depend on the underlying solidification pattern, this effect being particularly noticeable with the

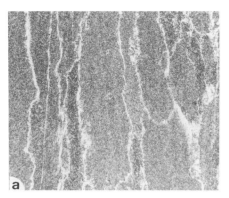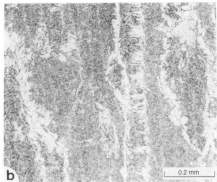

4.6 Typical as-deposited microstructure of 1.4%Mn welds deposited with interpass temperature of: *a*) 20 °C; *b*) 300 °C.

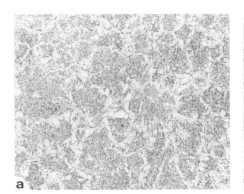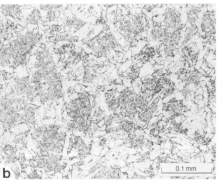

4.7 Typical coarse-grained microstructure of 1.4%Mn welds deposited with interpass temperature of: *a*) 20 °C; *b*) 300 °C.

weld of highest manganese content deposited with the highest inter-pass temperature.

The fine-grained reheated microstructures were also coarsened by increasing the interpass temperature, as shown in Fig. 4.8. Pearlite colonies were found with high interpass temperatures but were absent when they were lower. A linear relationship was found between the reciprocal of the square root of the mean intercept grain size and the interpass temperature, as illustrated in Fig. 4.9(*a*), although the lines for welds A and C were almost parallel to each other. Using yield

Table 4.3 Influence of interpass temperature on tensile properties for welds with varying manganese content

Interpass temp, °C	Electrode code	Yield stregnth, N/mm^2	Tensile strength, N/mm^2	Elongation, %	Reduction of area, %
20	A	450	510	29	78
	B	500	550	31	78
	C	540	600	27	75
	D	600	660	22	72
150	A	410	480	31	80
	B	450	520	29	79
	C	490	560	29	77
	D	520	600	30	77
240	A	390	460	35	81
	B	430	500	35	81
	C	450	520	32	79
	D	500	570	31	78
300	A	370	450	34	80
	B	410	490	35	81
	C	430	510	33	79
	D	450	530	32	81

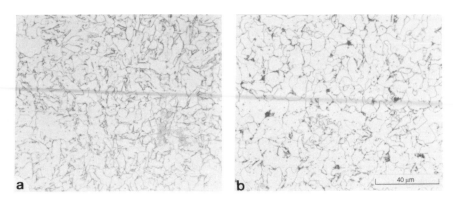

4.8 Typical fine-grained microstructure of 1.4%Mn welds deposited with interpass temperature of: *a*) 20 °C; *b*) 300 °C.

strength data from Table 4.3, linear relationships were also found between the grain size parameter of the fine grained region and the weld metal yield strength (Fig. 4.9(*b*)); here the two lines were also appreciably displaced.

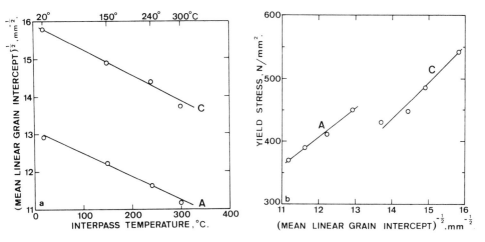

4.9 Effect of interpass temperature on: *a*) Ferrite grain size of refined regions; *b*) Weld metal yield strength.

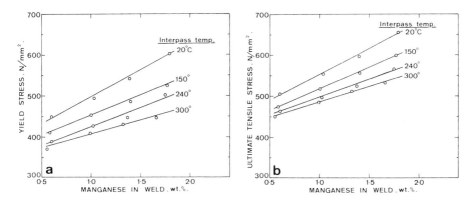

4.10 Effect of interpass temperature and manganese content on: *a*) Yield strength; *b*) Tensile strength.

Mechanical properties

The tensile properties of welds deposited with different interpass temperatures are given in Table 4.3. The linear relationships between manganese content and strength values for different temperatures are illustrated in Fig. 4.10; the best fit straight lines from which are given in Eq. [4.1]–[4.8], below. Increasing the interpass temperature reduced both strength values substantially, the decrease from 20 to

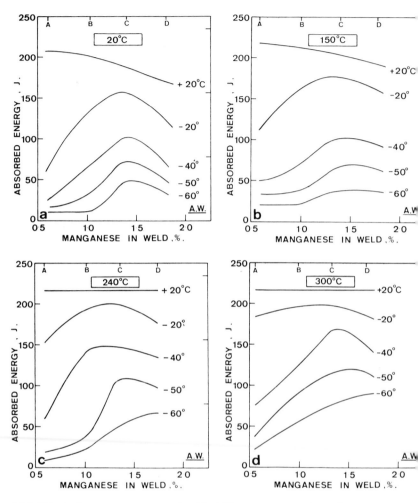

4.11 Effect of manganese on as-welded Charpy energy at varying temperature with interpass temperature of: a) 20 °C; b) 150 °C; c) 240 °C; d) 300 °C.

300 °C being about 100 N/mm^2 for the 1.4%Mn weld metal; ductility values increased slightly. In addition, increasing interpass temperature reduced the effect of manganese in increasing strength; the increase in both parameters fell from approximately 130 N/mm^2 per 1%Mn with 20 °C interpass to 70 N/mm^2 per 1%Mn with 300 °C.

The equations of the best fit lines are:

$$20\ ^\circ\text{C} \qquad \text{YS}_{aw} = 367 + 129\ \text{Mn} \qquad [4.1]$$

$$150\ ^\circ\text{C} \qquad \text{YS}_{aw} = 357 + 94\ \text{Mn} \qquad [4.2]$$

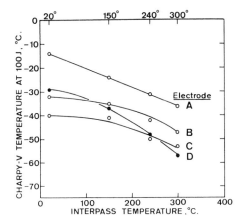

4.12 Effect of interpass temperature on 100 J Charpy transition temperature, as-welded.

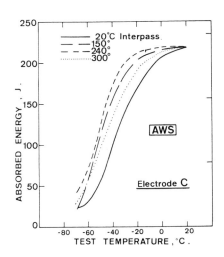

4.13 Effect of interpass temperature on as-welded Charpy transition curves for 1.4%Mn weld metal from AWS A 5.1–69 type deposits.

240 °C	$YS_{aw} = 331 + 93\ Mn$	[4.3]
300 °C	$YS_{aw} = 337 + 68\ Mn$	[4.4]
20 °C	$TS_{aw} = 426 + 126\ Mn$	[4.5]
150 °C	$TS_{aw} = 414 + 103\ Mn$	[4.6]
240 °C	$TS_{aw} = 411 + 86\ Mn$	[4.7]
300 °C	$TS_{aw} = 412 + 74\ Mn$	[4.8]

The results of Charpy impact tests are summarised in Fig. 4.11 and 4.12. As with the results discussed in the previous chapter, the optimum manganese content was again found to be about 1.4%, regardless of interpass temperature (Fig. 4.11). The 100 J transition temperature can be seen in Fig. 4.12 to decrease progressively for each electrode as interpass temperature was increased from 20 to 300 °C. This beneficial result is undoubtedly a result of the progressive reduction of both coarse as-deposited microstructure and yield strength of the deposits as the interpass temperature was raised. In AWS A 5.1–69 deposits which, as will be shown later in this chapter, consist almost entirely of fully recrystallised weld metal, the effect is absent and the beneficial effect of increasing the interpass temperature is reversed above 240 °C, as shown in Fig. 4.13 for the 1.4%Mn weld metal centreline.

The major influence of interpass temperature was to reduce the proportion of coarse columnar microstructure in an ISO 2560 deposit and this improved Charpy toughness transition temperatures, despite some coarsening of the recrystallised microstructure.

Electrode diameter

Tests to examine the influence of electrode diameter[4] were made on the four electrodes (coded A, B, C and D) of the different manganese levels examined previously – this time using core wires of 3.25, 4, 5 and 6 mm diameter. The coatings were formulated to maintain a manganese content independent of diameter and the coating factor was maintained constant at 1.68. Inevitably some differences had to be made to the operating conditions to fill the preparation of the ISO 2560:1973 testpiece. Welding was in the flat (downhand) position using DC positive polarity, a nominally constant welding (travel) speed and an interpass temperature of 200 °C. Other welding conditions are summarised in Table 4.4.

Only with the smallest electrode size did the number of beads per layer have to be increased from three to four to fill the preparation adequately. The disposition of runs and their resultant HAZs in the four welds are shown in Fig. 4.14. The 3.25 mm electrodes gave a deposit whose centreline region (i.e. that from which Charpy test specimen notches are machined) consisted largely of HAZ material, similar to the AWS A 5.1–69 deposit illustrated in Fig. 4.5.

Table 4.5 shows that increasing the electrode diameter through the range increased the heat input from 0.7 to 1.8 kJ/mm, whilst the weld

Table 4.4 Welding conditions for different electrode diameter

Electrode diameter, mm	Number of layers	Beads/ layer	Welding current, A	Arc volts, V	Travel speed, mm/sec
3.25	10	4	125	21	3.9
4	9	3	170	22	3.5
5	7	3	225	24	3.7
6	6	3	280	26	4.0

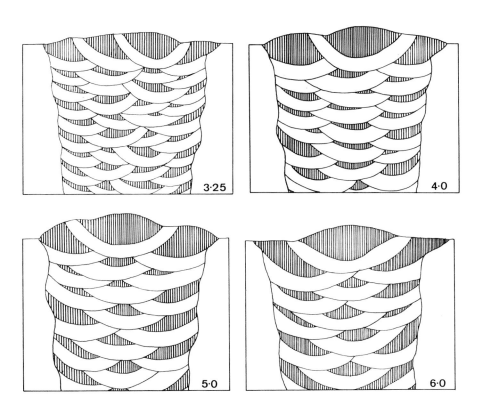

4.14 Cross sections of welds deposited with electrodes of different diameter (columnar weld metal hatched, recrystallised white).

cooling time from 800–500 °C (measured at the Institut de Soudure) doubled.

Chemical analysis (Table 4.6) showed that there was a slight tendency for manganese and silicon levels to be reduced as the electrode diameter was increased above 4 mm and heat input increased.

Table 4.5 Heat input and weld cooling with different
electrode diameter

Electrode diameter, mm	Heat input, kJ/mm	Cooling time, 800–500 °C, $T_{8/5}$, sec
3.25	0.7	5
4	1.1	7
5	1.5	9
6	1.8	10

Table 4.6 Influence of electrode diameter on composition
(wt%) of weld metal having varying manganese content

Electrode diameter, mm.	Electrode code	C	Mn	Si
3.25	A	0.041	0.56	0.25
	B	0.047	0.88	0.26
	C	0.047	1.29	0.27
	D	0.050	1.73	0.26
4	A	0.043	0.59	0.26
	B	0.056	0.94	0.27
	C	0.053	1.39	0.31
	D	0.059	1.80	0.32
5	A	0.033	0.55	0.25
	B	0.043	0.94	0.28
	C	0.048	1.34	0.28
	D	0.052	1.73	0.29
6	A	0.043	0.60	0.27
	B	0.044	0.85	0.26
	C	0.054	1.28	0.28
	D	0.054	1.62	0.27

Note: all welds contained 0.006–0.009%S and
0.008–0.015%P; typical levels of other elements are given
in Table 3.1.

Metallography

The macrostructures of welds deposited with electrodes of different
sizes, shown diagrammatically in Fig. 4.14, illustrate that, as the
electrode size was increased, fewer layers were required to complete
the weld. Figure 4.15 shows the rate at which both weld bead and
supercritical HAZ size increased as the electrode diameter was
increased. Linear relationships are apparent for both weld bead and

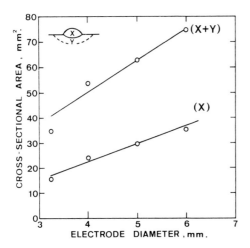

4.15 Effect of electrode size on weld bead (X) and recrystallised HAZ (Y) cross-sectional area.

HAZ, except for the smallest electrodes, which gave an unexpectedly small HAZ.

The distribution of the zones of the weld is illustrated for electrode C in Fig. 4.16. Figure 4.17 shows how the distribution for each region in the centre of the weld (the area in which the Charpy notches were located) varied with electrode size. Increasing electrode size increased the proportion of columnar weld metal at the expense of both coarse equiaxed and refined regions. The columnar grain width also increased with increasing electrode size.

Hardness measurements (HV5), taken down the centreline of different deposits, are summarised for the largest and smallest diameters of electrode C in Fig. 4.18 and related to the zonal structure (coded in the same way as in Fig. 4.14). The variation in hardness was more noticeable with 6 mm electrodes, amounting to a distinct waviness. There was also a slight fall in hardness up the weld, reaching a minimum in the fine grained region immediately below the final run.

Quantitative metallographic examination of the as-deposited microstructure at ×200 showed no significant change in the proportions of the constituents in the microstructure with electrode size (Fig. 4.19), except for electrode A, for which the proportion of grain boundary ferrite increased. However, the width of the grain boundary ferrite (Fig. 4.20) and the lath size of acicular ferrite (Fig. 4.21) both increased significantly as the electrode size was increased.

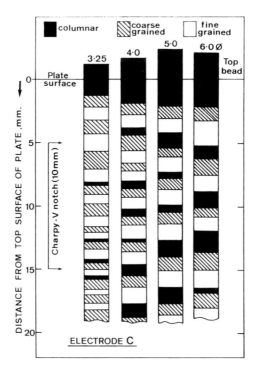

4.16 Influence of electrode size on zonal distribution at the Charpy notch position, electrode C.

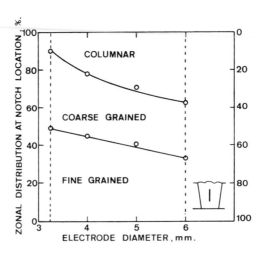

4.17 Zone distribution at the Charpy notch position, electrode C, for different electrode diameters.

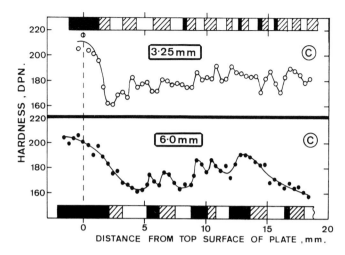

4.18 Hardness traverses (HV5), as-welded, down the centreline of welds deposited with electrode C of 3.25 and 6.0 mm diameter.

In the coarse grained reheated weld metal the width of grain boundary ferrite and the lath size of acicular ferrite increased with electrode size, as was the case in the as-deposited regions.

The fine grained weld metal also coarsened with increasing electrode diameter, as shown in Fig. 4.22 and 4.23. Careful examination showed that with small electrodes (Fig. 4.22(a)), the grains were duplex in character but with large electrodes, the structure was of ferrite grains with small pearlite colonies (Fig. 4.22(b)).

Mechanical properties

Hardness measurements on the final run of as-deposited weld metal (Table 4.7), showed relatively low values because of the high interpass temperature of 200 °C (cf. Table 4.2), with little perceptible influence of electrode diameter.

Tensile test results, detailed in Table 4.8 and plotted against manganese content in Fig. 4.24, showed little effect of electrode size, tensile strength falling by only 30 N/mm^2 over the whole range. Best fit straight lines of yield and tensile strength against manganese content (detailed in Ref. 4) all gave equations similar to Eq. [4.2] and [4.6] in the previous section, with 1%Mn increasing yield and tensile strength by about 100 N/mm^2.

Results from the present tests and those with variable interpass temperatures earlier in this chapter show a good linear relationship

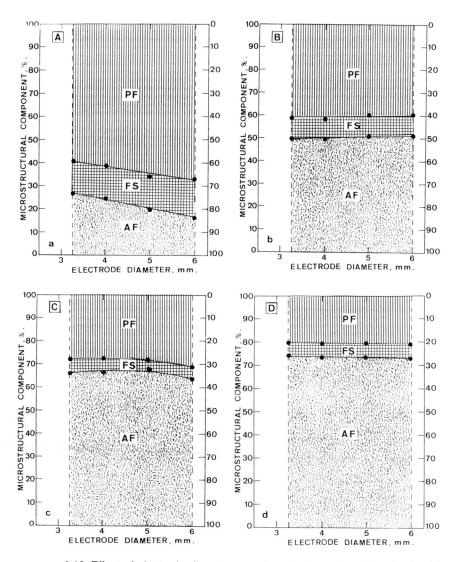

4.19 Effect of electrode diameter on microstructure of as-deposited weld metal of varying manganese level.

(Fig. 4.25) between the grain size of the fine-grained regions and yield strength. However, the present results are in marked contrast to earlier results, where heat input was altered by changing electrode diameter. Interpass temperature has a marked effect on weld strength, mainly because it decreases the proportion of as-deposited microstructure.

Table 4.7 Influence of electrode size on hardness of top
run beads having varying manganese content

Electrode dia, mm	3.25	4	5	6
Code	Hardness, HV5			
A	178	193	170	170
B	189	181	181	184
C	215	222	205	205
D	233	229	224	222

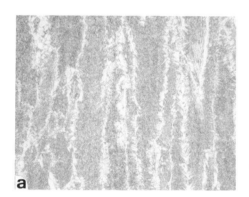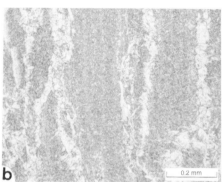

4.20 Microstructure of top bead of welds deposited with electrode C of: *a*)
3.25 mm; *b*) 6.0 mm diameter.

4.21 Microstructure of acicular ferrite in the top bead of welds deposited
with electrode C of: *a*) 3.25 mm; *b*) 6.0 mm diameter.

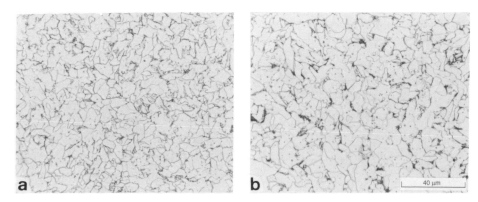

4.22 Microstructure of fine-grained weld metal deposited with electrode C of: *a*) 3.25 mm; *b*) 6.0 mm diameter.

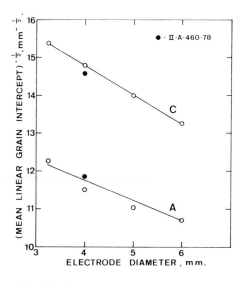

4.23 Effect of electrode diameter on mean grain size in the fine grained region of weld metal from electrode C. The solid circles represent results from tests in Fig. 4.9(*a*).

Although increasing electrode size also slows weld cooling, it has little effect on strength because it increases the proportion of stronger, as-deposited microstructure.

Charpy test results for the 3.25 mm electrode weld were not strictly comparable with the others because, as shown in Fig. 4.14 and 4.16,

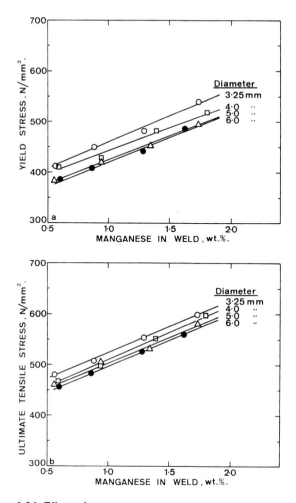

4.24 Effect of manganese content and electrode diameter on as-welded:
a) Yield strength; *b)* Tensile strength.

the notch region contained a higher proportion of refined microstructure. However, from the Charpy results for each weld, shown in Fig. 4.26, it is clear that increasing the size of electrode pushed the transition curves to higher temperatures and reduced the upper shelf energy. Figure 4.27 illustrates that, with different electrode diameters, the optimum manganese content for the best toughness remained close to 1.4%. The effect of electrode diameter can be seen in Fig. 4.28 to be most marked with the low manganese contents and was practically negligible at the highest level.

Table 4.8 Tensile test results on welds deposited with electrodes of different diameter

Electrode diameter, mm	Electrode code	Yield strength, N/mm²	Tensile strength, N/mm²	Elongation, %	Reduction of area, %
3.25	A	410	480	32	77
	B	450	510	31	80
	C	480	550	30	80
	D	540	600	29	77
4	A	410	470	32	81
	B	430	500	33	79
	C	480	550	31	78
	D	520	600	28	76
5	A	380	460	35	78
	B	420	500	36	79
	C	450	530	28	77
	D	500	580	29	75
6	A	390	460	34	80
	B	410	480	33	79
	C	440	530	29	78
	D	490	560	29	77

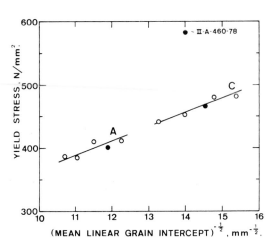

4.25 Relationship between as-welded yield strength and grain size of the fine grained region for deposits from electrodes A and C of different diameter. The solid circles represent results from tests in Fig. 4.9(*b*).

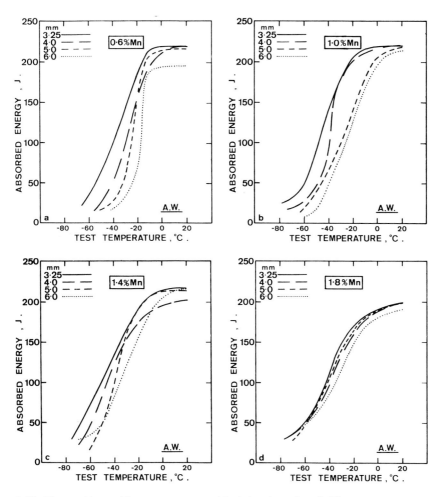

4.26 Charpy V transition curves, as-welded, for deposits of different manganese content made with different electrode diameters.

The detrimental effect of heavy gauge electrodes (and hence higher heat input) on toughness is in contrast to the influence of increasing interpass temperature (which also slows weld cooling). This is attributed to the heavier gauge electrodes increasing the proportion of coarse as-deposited microstructure in the deposit (Fig. 4.17) as well as to the coarser grain size of the refined regions (Fig. 4.22). In the welds where slow cooling was the result of high interpass temperatures, the coarser grain size of the refined regions was offset by the reduced proportion of coarse zones in the microstructure as the

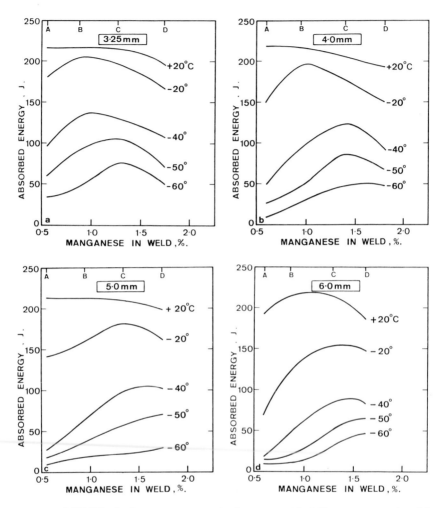

4.27 Effect of manganese content on as-welded Charpy energy level for welds deposited using different electrode diameters.

interpass temperature was increased. The combined influence of electrode diameter and interpass temperature on structure is sketched three-dimensionally in Fig. 4.29.

Heat input

Heat input varied in the previous section as a result of using electrodes of different diameters. The variation in this section[5] was pro-

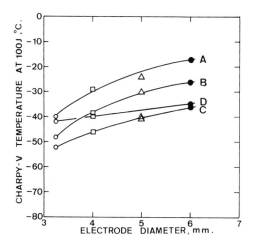

4.28 Effect of electrode diameter on temperature for 100 J Charpy energy for welds with varying manganese content, as-welded.

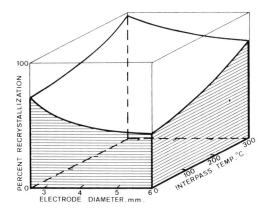

4.29 Showing combined effects of interpass temperature and electrode diameter on extent of recrystallisation for welds deposited using a three beads/layer sequence.

duced by welding with 4 mm electrodes of each manganese level at the welding speeds given in Table 4.9. As this resulted in different numbers of beads in each layer, the edges of the weld preparation were buttered with the electrodes under investigation to avoid differences in composition as a result of dilution of plate into the weld. They were then machined and the root gap varied from that in

Table 4.9 Welding conditions for different heat input

Beads per layer	Number of layers	Run out length,* mm	Welding, speed, mm/sec	Heat input, kJ/mm	800–500 °C cooling time, $T_{8/5}$, sec
4	12	660	6.2	0.6	4
3	9	400	3.6	1.0	7
2	6	200	1.7	2.2	13
1	6	100	0.87	4.3	34

Notes: * from 410 mm electrode length.
4 mm diameter electrodes were used in the flat position throughout with 170 A and 22 V arc voltage, electrode positive, and 200 °C maximum interpass temperature.

Table 4.10 Composition (wt%) of weld metal having varying manganese contents deposited at different heat input

Heat input, kJ/mm	Electrode code	C	Mn	Si
0.6	A	0.044	0.62	0.32
	B	0.046	0.96	0.31
	C	0.050	1.42	0.38
	D	0.055	1.93	0.35
1.0	A	0.037	0.60	0.29
	B	0.039	0.94	0.29
	C	0.048	1.41	0.35
	D	0.051	1.80	0.32
2.2	A	0.038	0.55	0.24
	B	0.036	0.89	0.24
	C	0.042	1.37	0.28
	D	0.045	1.69	0.26
4.3	A	0.043	0.52	0.20
	B	0.042	0.93	0.20
	C	0.043	1.37	0.24
	D	0.047	1.73	0.25

Note: all welds contained 0.006–0.009%S and 0.005–0.014%P; typical levels of other elements are given in Table 3.1.

ISO 2560:1973 to accommodate the number of beads in each layer. The welding technique varied from fine stringer beads for the lowest heat input to a full wide weave for the highest.

The chemical compositions, given in Table 4.10, show a slight tendency for the elements carbon, manganese and silicon to decrease

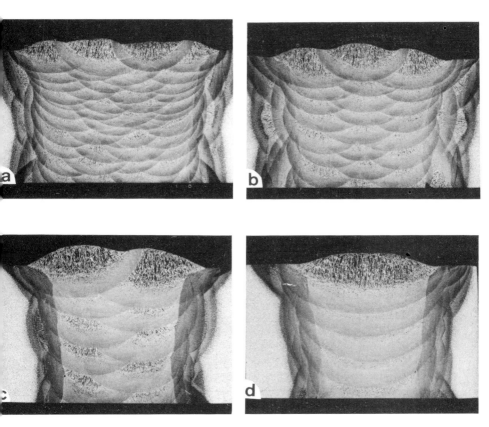

4.30 Typical macrosection of welds deposited with heat inputs of: *a*) 0.6; *b*) 1.0; *c*) 2.2; *d*) 4.3 kJ/mm.

with increasing heat input, as was later found for welds of varying phosphorus content, see Table 6.5.

Metallography

The disposition of the weld runs is shown in Fig. 4.30. Measurements on the top bead showed a linear dependence of both bead and HAZ size on heat input (Fig. 4.31). The different zones in the weld metal widened as the heat input was increased, and columnar microstructures were absent from the central regions of welds when the heat input was high.

To investigate this aspect further, sketches of actual and summed zone distributions near the centre line of the weld for welds with three

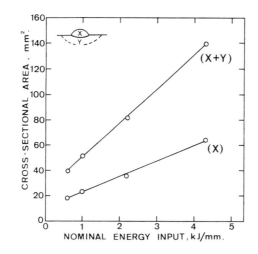

4.31 Influence of heat input on weld bead (X) and HAZ (Y) size.

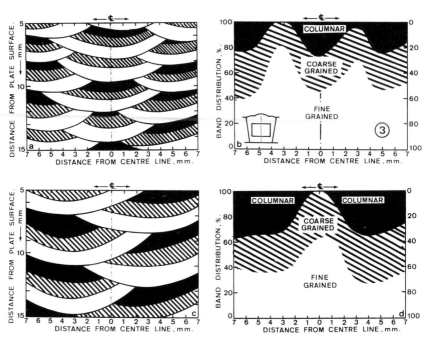

4.32 Actual (*a,c*) and summed (*b,d*) zonal distributions for welds with (*a,b*) three runs per layer and (*c,d*) two runs per layer.

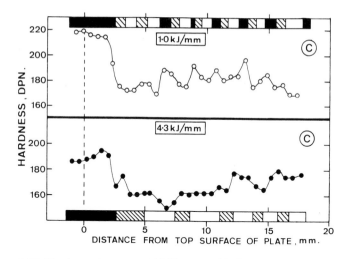

4.33 Hardness traverses (HV5), as-welded, down the centreline of welds deposited with electrode C at 1.0 and 4.3 kJ/mm.

and two passes per layer (1.0 and 2.2 kJ/mm, respectively) were prepared (Fig. 4.32). These sketches show that, with three beads per layer, the proportion of columnar structure was at a maximum and that of fine grained at a minimum in the central region. The opposite is true for the weld with two runs per layer, which is similar to the AWS A 5.1–69 deposit shown later in Fig. 4.54. With full weaving, no columnar structure remained in the 4.3 kJ/mm deposit, despite the weld centreline being in the optimum position for such a structure. This fully weaved deposit gave a less wave-like trace (and a lower hardness level) than the 1 kJ/mm deposit when hardness values down the centreline of the weld were plotted (Fig. 4.33).

In the columnar region, the columnar grain width increased in proportion to the heat input. Point counting showed that heat input increased the proportion of both grain boundary ferrite and ferrite with aligned second phase at the expense of acicular ferrite (Fig. 4.34), although the effect was not large. More significant was the influence of heat input on the coarseness of the microstructure; this can be seen for the columnar zone in Fig. 4.35 and 4.36. The coarse-grained region (Fig. 4.37), developed ferrite with aligned second phase to a noticeable degree. Figure 4.38 shows how the grain size of the fine grained region increased with heat input. As with the welds made with different sized electrodes (Fig. 4.22), the fine grains with the lowest heat input exhibited a duplex structure, whilst those with the highest heat input contained some fine pearlite.

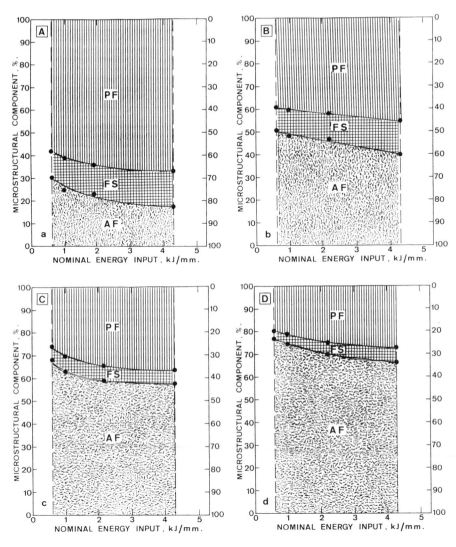

4.34 Effect of heat input on the microstructure of as-deposited weld metal containing a nominal: *a*) 0.6%; *b*) 1.0%; *c*) 1.4%; *d*) 1.8%Mn.

Mechanical properties

Hardness of the top beads, summarised in Table 4.11, shows a decrease averaging 25 HV5 over the full range of heat inputs; values were comparatively low because of the high interpass temperature.

Tensile test results, summarised in Table 4.12 and Fig. 4.39, showed a marked drop in strength as the heat input was increased. The drop

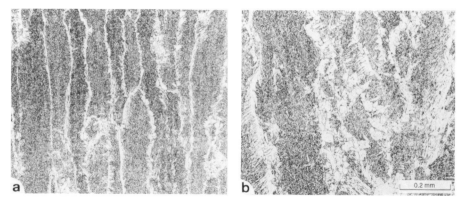

4.35 Microstructure of top bead of welds deposited from electrode C with heat inputs of: *a*) 0.6; *b*) 4.3 kJ/mm.

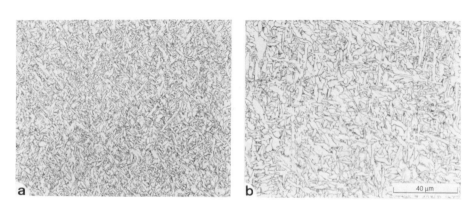

4.36 Microstructure of acicular ferrite in the top bead of welds deposited from electrode C with heat inputs of: *a*) 0.6; *b*) 4.3 kJ/mm.

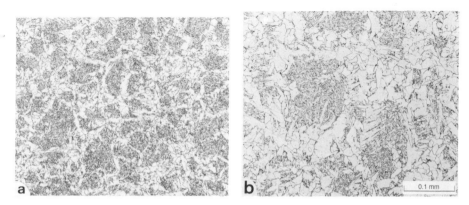

4.37 Microstructure of coarse-grained electrode C weld metal deposited with heat input of: *a*) 0.6; *b*) 4.3 kJ/mm.

Table 4.11 Influence of heat input on hardness of top run beads having varying manganese content

Electrode code	A	B	C	D
Heat input, kJ/mm	Hardness, HV5			
0.6	176	191	223	233
1.0	169	183	215	226
2.2	171	178	203	228
4.3	155	175	184	208

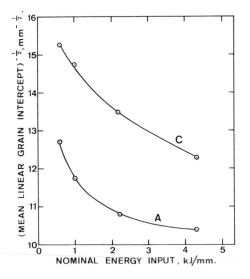

4.38 Effect of heat input on mean grain size in the fine-grained region of weld metal from electrodes A and C.

in yield strength, approximately 100 N/mm² at low manganese contents, increased slightly as the manganese content was increased. The decrease in tensile strength was smaller (less than 50 N/mm²) at low manganese levels but increased to nearly 100 N/mm² at the highest manganese level. Equations relating strength to manganese content at different heat inputs were determined as follows:

$$0.6 \text{ kJ/mm} \quad YS_{aw} = 376 + 105 \text{ Mn} \quad [4.9]$$

$$1.0 \text{ kJ/mm} \quad YS_{aw} = 352 + 88 \text{ Mn} \quad [4.10]$$

$$2.2 \text{ kJ/mm} \quad YS_{aw} = 351 + 62 \text{ Mn} \quad [4.11]$$

$$4.3 \text{ kJ/mm} \quad YS_{aw} = 301 + 78 \text{ Mn} \quad [4.12]$$

Table 4.12 Tensile test results for welds of varying manganese contents deposited with different heat input

Heat input, kJ/mm	Electrode code	Yield strength, N/mm^2	Tensile strength, N/mm^2	Elongation, %	Reduction of area, %
0.6	A	440	500	31	80
	B	480	530	28	77
	C	510	580	27	79
	D	590	630	24	77
1.0	A	400	470	32	82
	B	440	500	33	81
	C	480	550	30	79
	D	510	590	29	77
2.2	A	390	460	30	81
	B	400	480	33	81
	C	440	520	30	82
	D	460	550	32	79
4.3	A	340	450	33	79
	B	380	480	33	79
	C	400	500	32	81
	D	440	540	30	79

$$0.6 \text{ kJ/mm} \qquad TS_{aw} = 428 + 105 \text{ Mn} \qquad [4.13]$$

$$1.0 \text{ kJ/mm} \qquad TS_{aw} = 406 + 102 \text{ Mn} \qquad [4.14]$$

$$2.2 \text{ kJ/mm} \qquad TS_{aw} = 410 + 80 \text{ Mn} \qquad [4.15]$$

$$4.3 \text{ kJ/mm} \qquad TS_{aw} = 417 + 68 \text{ Mn} \qquad [4.16]$$

These show the decreasing influence of manganese as heat input was increased. This, like the earlier effect of interpass temperature (Eq. [4.1]–[4.8]), is undoubtedly a result of the increasing heat input (or interpass temperature) decreasing the cooling rate so that less acicular ferrite formed for a given increase in manganese.

When grain size in the fine grained region was plotted against yield strength, two separate curves were obtained for electrodes A and C (Fig. 4.40). The separation of the two curves was less than in the case of the welds made with different interpass temperature (Fig. 4.9(*b*)) but greater than with electrodes of different diameter, where a single line was obtained (Fig. 4.25).

From Charpy transition curves, plots of manganese content against Charpy energy at different heat inputs (Fig 4.41) show that the optimum manganese content was maintained close to 1.4%. Figure 4.42 shows that the optimum heat input appears to be 2.2 kJ/mm. However, the position of the Charpy notch in relation to the number of runs has some bearing on this finding, as the 2.2 kJ/mm weld was

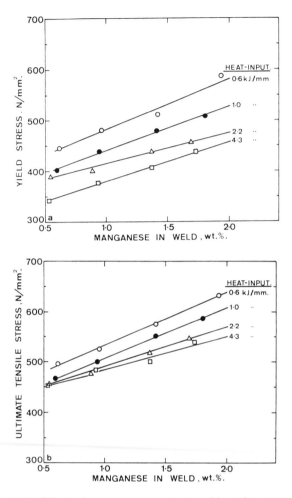

4.39 Effect of manganese content and heat input on as-welded: *a)* Yield strength; *b)* Tensile strength.

one where the notch was inevitably located in a predominantly fine grained region where two runs overlapped (Fig. 4.30).

Additional tests were, therefore, carried out in which the Charpy notches of 1.0 and 2.2 kJ/mm welds were made at the weld centre-line and also displaced about 3 mm to one side, so that the 2.2 kJ/mm notches originated in weld metal containing columnar structure and the 1.0 kJ/mm notches were in a bead overlap region. The results, summarised in Table 4.13, show not only a markedly higher transition temperature for each weld when notched in a region containing

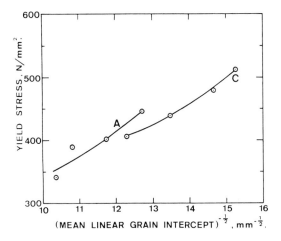

4.40 Relationship between as-welded yield strength and grain size of the fine grained region for deposits of different heat input from electrodes A and C.

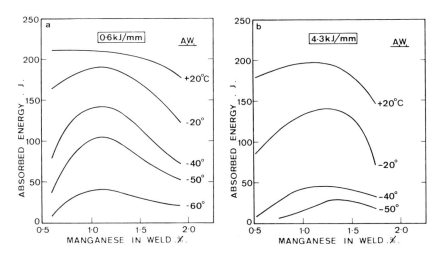

4.41 Effect of manganese content on as-welded Charpy energy level for welds deposited with heat input of: *a*) 0.6; *b*) 4.3 kJ/mm.

columnar structure, but the comparative difference between the two heat inputs was reversed when like was compared with like. The lower heat input always gave the lower transition temperature when comparison was made with notch locations in the same type of weld microstructure.

Table 4.13 Effect of notch location on Charpy results

Heat input, kJ/mm	Electrode code	Temperature, °C, for 100 J Charpy energy		Difference, °C
		Centreline notch	Displaced notch	
1.0	A	−31	−50	−19
	B	−46	−66	−20
	C	−54	−63	−9
	D	−44	−56	−12
2.2	A	−39	−24	+15
	B	−46	−30	+16
	C	−57	−46	+11
	D	−38	−37	+1

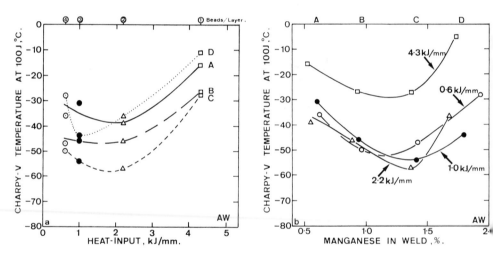

4.42 Effect on temperature for 100 J Charpy energy, as-welded of: *a)* Heat input; *b)* Manganese content.

Welding position

This part of the investigation[6] was restricted to a comparison of flat and vertical-up welding with full and limited weaving. Because of the wide weave, the plates were buttered with the appropriate consumables before welding. Batches of 4 mm diameter electrodes, coded A–D, were used with an interpass temperature of 200 °C and testing was carried out in the as-welded condition, with tensile testing after hydrogen removal. Welding conditions are detailed in Table 4.14.

Table 4.14 Welding conditions for different positional welding techniques

Welding position	Beads per layer	Number of layers	Run out length*, mm	Welding current, A	Arc voltage, V	Welding, speed, mm/sec
Flat	2	6	200	170	22	1.7
	1	6	100	170	22	0.9
Vertical	2	4	110	135	20	0.8
	1	5	90	135	20	0.7

Notes: *from 410 mm electrode length.
4 mm diameter electrodes were used with 200 °C maximum interpass temperature.

Table 4.15 Parameters derived from positional welding conditions

Welding position	Beads per layer	Heat input, kJ/mm	800–500 °C cooling time, $T_{8/5}$, sec
Flat	2	2.2	13
	1	4.3	34
Vertical	2	3.4	22
	1	3.9	30

Table 4.16 Composition (wt%) of weld metal having varying manganese content, deposited in different welding positions at different heat inputs

Welding position	Beads/ layer	Electrode code	C	Mn	Si	N	O
Flat	2	A	0.038	0.55	0.24	0.011	0.056
		B	0.036	0.89	0.24	0.008	0.051
		C	0.042	1.37	0.28	0.013	0.047
		D	0.045	1.69	0.26	0.015	0.048
Flat	1	A	0.043	0.52	0.20	0.014	0.047
		B	0.042	0.93	0.20	0.015	0.047
		C	0.043	1.37	0.24	0.014	0.042
		D	0.047	1.73	0.25	0.015	0.037
Vertical	2	A	0.046	0.60	0.28	0.007	0.056
		B	0.052	0.97	0.32	0.008	0.048
		C	0.055	1.38	0.29	0.008	0.048
		D	0.071	1.85	0.33	0.008	0.050
Vertical	1	A	0.040	0.57	0.24	0.006	0.062
		B	0.048	0.98	0.26	0.012	0.060
		C	0.050	1.28	0.26	0.008	0.048
		D	0.057	1.74	0.27	0.010	0.048

Note: all welds contained 0.007–0.009%S and 0.005–0.016%P; typical levels of other elements are given in Table 3.1.

Values derived from the primary welding parameters, including cooling time measurements carried out by the Institut de Soudure, are included in Table 4.15. These show that the cooling time increased as the heat input was increased, although no other obvious relations are apparent. The lowest heat input and the fastest weld cooling were found with the limited weave flat weld and the highest heat input and the slowest cooling with the fully weaved flat weld; there was much less difference between the two vertical welds.

The chemical compositions of the welds, detailed in Table 4.16, show that the vertical welds contained marginally more carbon, manganese, silicon and oxygen, but less nitrogen than those deposited in the flat position. There were no consistent differences between wide and narrow weave; in fact, some apparent trends in the flat position were reversed in the vertical.

Metallography

Macrosections of the test welds are shown in Fig. 4.43. Fewer runs were required in vertical than in flat welding and the centreline of the limited weave welds contained little as-deposited structure, especially in the flat welds. Dimensions of top beads and their underlying HAZs, detailed in Table 4.17, show that the cross-sectional areas of the weld beads did not increase in the same order as the heat inputs, although the HAZ areas did. This apparent anomaly is likely to be a result of the difference in the mechanism of weld metal deposition between the two techniques. That such a difference exists is apparent in the weld metal sections in Fig. 4.43, where increased penetration (which is not apparent in the other deposits) can be seen clearly at the extremities of the fully weaved vertical weld.

Zonal distributions in the neighbourhood of the Charpy notch, together with the summed distribution in the same area, are shown in Fig. 4.44. Neither flat weld contained any columnar structure at or near the centreline, the full weave being fully recrystallised. Both vertical welds, however, contained appreciable amounts of columnar structure, although only a small amount was present in the two-bead weld where the beads joined, but this was not at the weld centreline.

There was little difference between the techniques when the hardness traverses along the weld centreline were compared, although the vertical weld traverses were less uniform as a result of the presence of the columnar structure.

In transverse sections (Fig. 4.43) the columnar grains of the flat welds appeared to have a greater length to width ratio than those of the vertical welds. However, examination of longitudinal sections showed that the grains in the vertical welds were flatter, being inclined

Table 4.17 Influence of welding position on weld bead and HAZ size

Welding position	Beads per layer	Heat input, kJ/mm	Bead area, X, mm²	Bead + HAZ area, X + Y, mm²	HAZ area, Y, mm²
Flat	2	2.2	36	82	46
	1	4.3	64	140	76
Vertical	2	3.4	68	124	56
	1	3.9	89	155	66

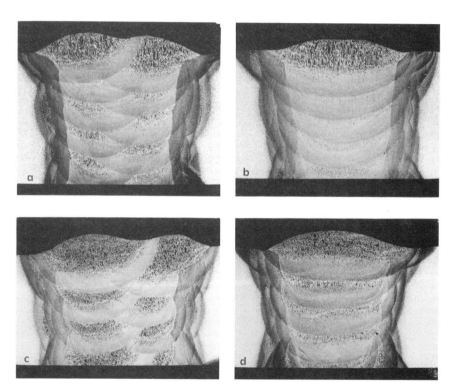

4.43 Typical macrosections of welds deposited: *a)* Flat with limited weave; *b)* Flat with full weave; *c)* Vertical-up with limited weave; *d)* Vertical-up with full weave.

at 66° to the weld surface compared with 77° for those deposited in the horizontal position. The weaved vertical welds were also found to have wider grains in the dwell positions at the ends of the weave than elsewhere.

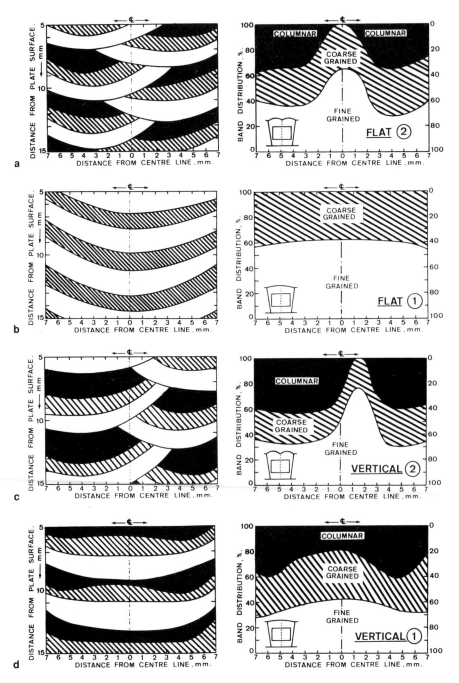

4.44 Actual (left) and summed (right) zonal distributions for welds deposited with electrode B using flat welding with: *a*) Two beads; *b*) Flat with full weave; *c*) Vertical with two beads; *d*) Vertical with full weave.

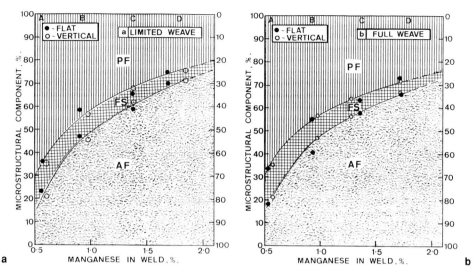

4.45 Influence of manganese content on microstructure of welds made with: *a*) Restricted weave; *b*) Full weave.

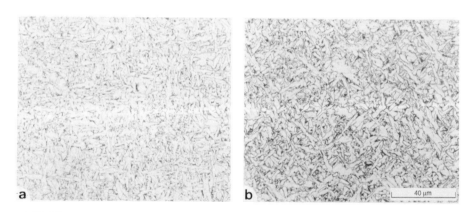

4.46 Microstructure of acicular ferrite in the top bead of welds deposited: *a*) Horizontally; *b*) Vertically.

Quantitative microscopy of the top beads (Fig. 4.45), showed that at all manganese levels there was no difference between the proportions of constituents of welds made horizontally and vertically, although the fully weaved welds contained marginally more grain boundary ferrite than those with a limited weave. From a careful

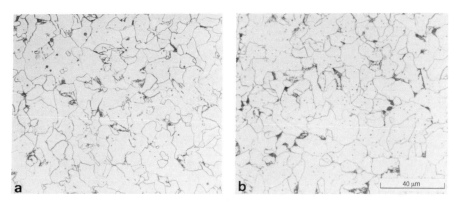

4.47 Microstructure of fine-grained weld metal deposited with a full weave: *a)* In the flat position; *b)* Vertically, ×630.

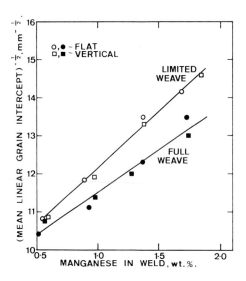

4.48 Effect of welding technique and manganese content on mean grain size in fine-grained weld metal.

examination, it appeared that the acicular ferrite grains (Fig. 4.46) were slightly finer in welds deposited in the flat position than in the vertical, as would be expected from the slightly lower heat inputs and faster cooling (Table 4.15).

The high temperature reheated weld metal was similar for both limited weave welds but with the full weave, the ferrite envelopes

Table 4.18 Influence of heat input on hardness of top run beads having varying manganese contents and deposited with different techniques

Welding technique	F, 2 runs	V, 2 runs	F, 1 run	V, 1 run
Electrode code	Hardness, HV5			
A	171	170	155	151
B	178	188	175	185
C	203	195	184	188
D	228	224	208	205

Note: F – flat; V – vertical; 1 run – full weave; 2 runs – limited weave.

round the finer transformation products were coarser near the weld centreline position for the vertical welds than for the flat position. Fine-grained weld metal (Fig. 4.47) contained some pearlite on account of its slow cooling. Grain size measurements (Fig. 4.48) showed separate trends for limited and full weaved deposits but no difference was apparent between welds deposited in the flat and vertical positions.

Mechanical properties

Hardness measurements on the top, untempered beads (Table 4.18) showed no significant differences between flat and vertical welding, although the full weave welds had slightly lower hardness than those deposited with a limited weave.

From the tensile test results detailed in Table 4.19, the influence of manganese on strength values with each different welding technique have been plotted in Fig. 4.49. The best fit straight lines from these graphs are:

2F	$YS_{aw} = 351 + 62\ Mn$	[4.17]
2V	$YS_{aw} = 326 + 81\ Mn$	[4.18]
1F	$YS_{aw} = 301 + 78\ Mn$	[4.19]
1V	$YS_{aw} = 275 + 86\ Mn$	[4.20]
2F	$TS_{sr} = 410 + 80\ Mn$	[4.21]
2V	$TS_{sr} = 413 + 88\ Mn$	[4.22]
1F	$TS_{sr} = 417 + 68\ Mn$	[4.23]
1V	$TS_{sr} = 317 + 100\ Mn$	[4.24]

The steeper slopes of the vertical welds may be a result of the carbon and perhaps silicon contents increasing slightly more with increasing

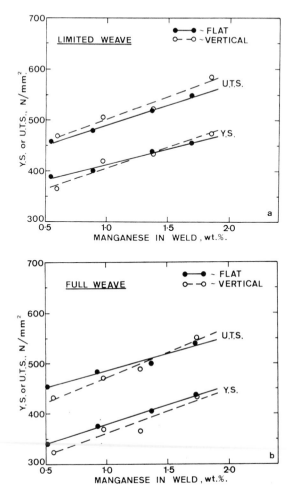

4.49 Effect of welding technique and manganese content on as-welded yield strength and tensile strength for: *a*) Limited weave; *b*) Fully weaved welds.

manganese content in the vertical welds than in the flat welds (Table 4.16). It may also be because the vertical welds contain more columnar microstructure in the central regions (Fig. 4.44).

A plot of the relation between yield strength and the grain size in the fine grained region (Fig. 4.50), appears to show a single relationship with a reasonable degree of scatter, in view of the range of compositions and the different degrees of recrystallisation in the deposits considered.

Table 4.19 Tensile test results for welds of varying manganese content deposited with different welding techniques

Welding technique	Electrode code	Yield strength, N/mm²	Tensile strength, N/mm²	Elongation, %	Reduction of area, %
F, 2 runs	A	390	460	30	81
	B	400	480	33	81
	C	440	520	30	82
	D	460	550	32	79
V, 2 runs	A	360	470	31	79
	B	420	500	30	79
	C	430	520	31	79
	D	470	580	26	75
F, 1 run	A	340	450	33	79
	B	380	480	33	79
	C	400	500	32	81
	D	440	540	30	79
V, 1 run	A	320	430	32	79
	B	370	470	31	79
	C	370	490	30	79
	D	430	550	27	77

Note: F – flat; V vertical; 1 run – full weave; 2 runs – limited weave.

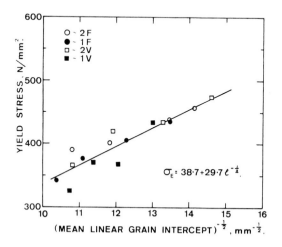

4.50 Relationship between as-welded yield strength and grain size of the fine grained region for deposits from electrodes A and C made using different welding techniques.

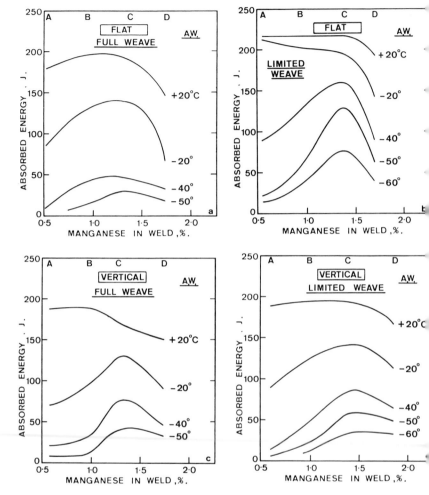

4.51 Effect of manganese content on Charpy energy level for welds deposited with different welding techniques, as-welded.

The Charpy data, summarised in Fig. 4.51 and 4.52, show a marked increase in transition temperature as the weave width was increased for the flat welds. However, this increase was only found for the vertical welds of the highest manganese content. The relative levels of toughness are in line with the differences in heat input and weld cooling seen in Table 4.15. Generally the optimum manganese content appears to be close to 1.4%.

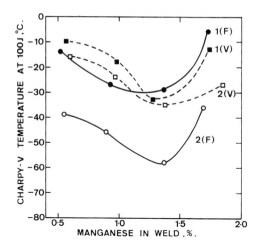

4.52 Effect of heat input and welding technique on temperature for 100 J Charpy energy for welds with varying manganese content, as-welded.

Testing specifications

The essential requirements of the AWS A 5.1–69 standard are shown in Fig. 4.53 and can be compared with the ISO 2560–1973 specimen details previously given in Fig. 2.2. The significant differences were that for A 5.1–69 the preparation was narrower at the root (12.7 mm as against 16 mm) and that weaving, rather than a stringer bead technique, was used. The changes resulted, for the American testpiece, in a higher heat input, which increased as welding progressed. These features led to differences between the structures and properties of the two deposits, which are examined below.

Since 1973, both specifications have changed somewhat; ISO 2560 now specifies two beads/layer and AWS A 5.1 now has a steeper preparation, bringing it closer to the ISO standard and, therefore, requiring less change in welding technique as welding progresses. It should be pointed out that the earlier ISO 2560:1973 did not specify the number of beads/layer and three were selected for the present programme to provide realistic weld metals containing a certain proportion of columnar microstructure for testing.

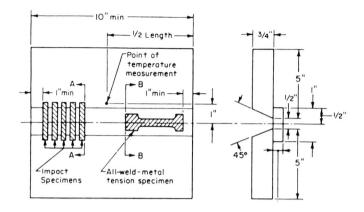

(a) Test Plate Showing Location of Test Specimens

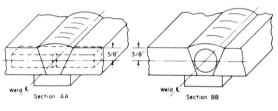

(b) Orientation of Impact Specimen   (c) Location of All-Weld-Metal Tension Specimen

4.53 Details of AWS test assembly (1″ [inch] = 25.4 mm): a) Test plate showing location of test specimens; b) Orientation of impact specimen; c) Location of all-weld-metal tension specimen.

Table 4.20 Weld metal composition with different test assemblies

Element, wt%		C	Mn	Si
Type of test	Electrode code			
ISO	A	0.035	0.66	0.30
2560:1973	B	0.038	1.00	0.30
	C	0.049	1.42	0.34
	D	0.051	1.82	0.34
AWS	A	0.031	0.56	0.23
A5.1-69	B	0.037	0.91	0.27
	C	0.040	1.32	0.28
	D	0.044	1.70	0.29

Note: all welds contained 0.005–0.008%S and 0.011–0.017%P; typical levels of other elements are given in Table 3.1.

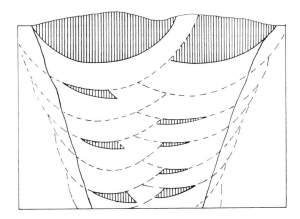

4.54 Cross section of weld deposited with AWS A5.1–69 test assembly (columnar weld metal hatched, recrystallised white).

In the comparative tests,[7] 4 mm diameter electrodes of different manganese contents (electrodes A–D) were used. With the ISO test assembly, the heat input employed was 1 kJ/mm, using a stringer bead technique. With the AWS A 5.1, the weaving technique employed gave twice the heat input – 2 kJ/mm. Other details, including an interpass temperature of 150 °C and 14 hour hydrogen removal treatment at 250 °C were identical. Tests were confined to chemical analysis, macro examination, and tensile and Charpy testing in the as-welded condition.

Chemical analyses (Table 4.20), showed slightly, but significantly, lower levels of C, Mn and Si in the AWS A 5.1 deposits, as a result of the higher heat input and weaving.

Metallography

Macroscopic examination confirmed that twice as many runs were required to deposit the ISO 2560 test assemblies as were needed for the AWS A 5.1; this is illustrated diagrammatically in Fig. 4.54, which should be compared with Fig. 2.1. The illustration also shows that the AWS A 5.1–69 assembly gives almost complete refinement of the weld microstructure down the central regions of the weld, the region from which most of each tensile test specimen and the notch of each Charpy specimen originates. Of all the sections examined, only one AWS A 5.1 specimen was found to contain a small tongue of unrefined weld metal in the central region. Examination described elsewhere, in this and other chapters, showed that, with the conditions used, about 20–25% unrefined weld metal can be expected in an ISO 2560 deposit with three beads/layer.

Mechanical properties

The results of tensile tests, summarised in Table 4.21, show that the AWS A 5.1 assembly resulted in lower strength and higher ductility values than ISO 2560. Regression analysis to show the effect of manganese gave the following equations for the best fit straight lines:

$$\text{ISO 2560} \qquad YS_{aw} = 314 + 108 \text{ Mn} \qquad [4.25]$$

$$\text{Assembly} \qquad TS_{aw} = 394 + 108 \text{ Mn} \qquad [4.26]$$

$$\text{AWS A 5.1} \qquad YS_{aw} = 332 + 75 \text{ Mn} \qquad [4.27]$$

$$\text{Assembly} \qquad TS_{aw} = 411 + 72 \text{ Mn} \qquad [4.28]$$

The differences in strength increased from almost zero at 0.6%Mn to 25 N/mm^2 at 1.25%Mn (the previous limit for the AWS A 5.1–69 E7018 basic electrode range – it is now 1.6%) and to nearly 50 N/mm^2 at 1.8%Mn. These differences were largely due to the reduced strengthening effect of manganese in the AWS test. As was apparent in Chapter 3, the effect of manganese on weld metal strength in the ISO test assembly is enhanced because of its role in promoting acicular ferrite in the columnar regions. Such regions are virtually absent from the central regions of AWS A 5.1–69 deposit weld metal, so that the enhanced strengthening effect of manganese is lost, as can be seen in Fig. 4.55.

Table 4.21 Tensile test results from different test assemblies

Test assembly	Electrode code	Yield strength, N/mm^2	Tensile strength, N/mm^2	Elongation, %	Reduction of area, %
ISO	A	390	470	32	81
25460:1973	B	410	500	31	81
	C	470	550	29	79
	D	510	590	28	77
AWS	A	380	460	33	81
A5.1-69	B	400	480	32	81
	C	420	500	34	81
	D	460	540	30	81

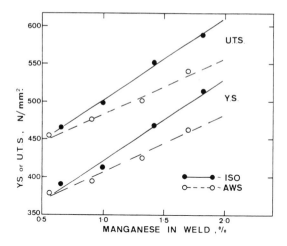

4.55 Effect of manganese content on as-welded strength properties with ISO 2560:1973 and AWS A5.1–69 test assemblies.

Table 4.22 Temperatures for 100 J Charpy energy with different test assemblies

Electrode	Temperature, °C, for 100 J		Difference, °C
	ISO 2560:1973	AWS A5.1	
A	−27	−37	−10
B	−44	−49	−5
C	−53	−53	0
D	−43	−43	0

Recent comparative tests[8] on low alloy steel deposits using both ISO 2560:1973 and AWS A 5.1–69 assemblies have also shown similar differences to those found here at the high manganese level.

Charpy impact tests showed generally similar transition curves and test values for the different test assemblies (Table 4.22). However, because of the differences in manganese level found with the two test assemblies, the optimum manganese content was lower when using the AWS A 5.1 method (1.3%Mn) than with the ISO 2560:1973

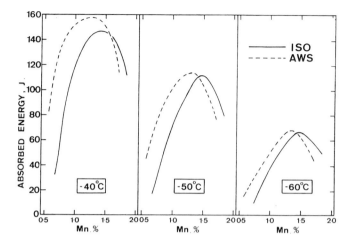

4.56 Effect of manganese content on as-welded Charpy energy at different temperatures with AWS A5.1–69 and ISO 2560:1973 test assemblies.

technique (1.5%Mn) (Fig. 4.56). This figure also shows that the AWS A 5.1 assembly tended to give higher upper shelf values at the optimum manganese level, although at lower temperatures, and hence lower maxima, the values were similar.

Increasing either interpass temperature or heat input (both of which are known to slow weld cooling, see elsewhere in this chapter) had no detectable effect on optimum manganese level using the ISO test assembly. However, changing from ISO 2560:1973 to AWS A 5.1 (which also increased heat input and slowed weld cooling) reduced the optimum manganese level by ~0.2%. This is evidently a consequence of the test region of the AWS deposits being devoid of any columnar microstructure, so that the beneficial influence of acicular ferrite cannot be made manifest.

Discussion

The tests described here have built on and reinforced those in the previous chapter and confirmed that the manganese content giving the best balance of strength and toughness is close to 1.4% over a wide range of manual welding conditions. It has also brought out some factors of major importance in both testing and use of electrodes for manual welding.

In most practical welding, it is difficult to ensure the absence of as-deposited columnar microstructure. It is this structure which, in many cases, gives the greatest variation in hardness and probably the lowest level of toughness, particularly if searching tests such as CTOD are employed, as these (like some real life failures) are capable of finding, and giving crack extension from, small brittle regions. In comparing the two types of test examined, the ISO 2560:1973 method, with three beads per layer, is to be preferred to the AWS A 5.1–69 with two, as it avoids a completely refined region down the centreline of the weld. Too high an interpass temperature is also to be deprecated as it will reduce the columnar zones to almost nothing; it may also give problems in achieving adequate strength. Unfortunately, the current revision of ISO 2560 specifies two beads/layer. This makes it less useful for assessing the practical utility of any electrodes tested as, like AWS A 5.1, it must give a testing region which is too refined to represent a normal fabrication weld.

Sufficient data were amassed to examine the influence of welding variables on properties over a useful range. An excellent correlation was obtained between the 800–500 °C cooling time ($T_{8/5}$) and heat inputs of 1 kJ/mm and above, provided that the interpass temperature was held constant (Fig. 4.57). With 200 °C interpass temperature, the logarithm of the cooling time and the heat input (HI) at heat inputs of 1.0 kJ/mm and greater were linearly related:

$$\mathrm{Log}\left(T_{8/5}\right) = 0.626 + 0.214 \; \mathrm{HI} \qquad\qquad [4.29]$$

At low values of heat input, shorter cooling times were measured, presumably because of the onset of three-dimensional cooling. Figure 4.57 also gives some idea of the marked influence of interpass temperature on the relationship at 1 kJ/mm.

The effect of interpass temperature is also apparent in the relationship between weld metal properties. The relation between weld cooling time and hardness is approximately linear for welds deposited with a constant interpass temperature of 200 °C. Relationships between weld cooling (or heat input) and yield strength were not linear, the greatest change in strength occurring at low heat inputs.

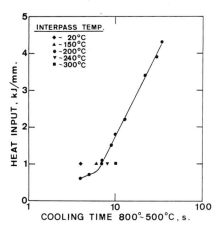

4.57 Relation between heat input and weld cooling time.

References

1 Evans GM: 'Effect of interpass temperature on the microstructure and properties of C–Mn all-weld-metal deposits'. *Welding Review* 1982 **1**(1) 14–20; *Welding Res Abroad* 1983 **29**(1) 13–23; IIW Doc II–A–460–78; *Schweissmitteilungen* 1979 **37**(87) 17–31.

2 Jones RL: 'Development of two-layer deposition techniques for the MMA repair welding of thick C–Mn steel plate without PWHT'. TWI Res Report 335/1987, TWI, Cambridge, April 1987.

3 Alberry PJ, Myers J and Chew B: 'An improved welding technique for HAZ refinement'. *Weld Metal Fabrication* 1977 **45** 549–53.

4 Evans GM: 'Effect of electrode diameter on the microstructure and properties of C–Mn all-weld-metal deposits'. *Welding Review* 1982 **1**(2) 4–8; *Welding Res Abroad* 1983 **29**(1) 24–34; IIW Doc II–A–469–79; *Schweissmitteilungen* 1980 **38**(90) 4–17.

5 Evans GM: 'Effect of heat input on the microstructure and properties of C–Mn all-weld-metal deposits'. *Welding J* 1982 **61**(4) 125s–32s; *Welding Res Abroad* 1983 **29**(1) 35–45; IIW Doc II–A–490–79; *Schweissmitteilungen* 1980 **38**(92) 20–35.

6 Evans GM: 'Effect of welding position on the microstructure and properties of C–Mn all-weld-metal deposits'. *Welding Review* 1982 **1**(3) 6–10; *Welding Res Abroad* 1983 **29**(1) 46–57; *Schweissmitteilungen* 1980 **38**(93) 12–27.

7 Evans GM: 'ISO/AWS; Comparison of ISO 2560 and AWS A 5.1–69'. *Welding Res Abroad* 1992 **38**(8/9) 45–9; *Schweissmitteilungen* 1978 **36**(85) 4–10.

8 Johnson MQ, Evans GM and Edwards GR: 'The influence of titanium additions and interpass temperature on the microstructure and mechanical properties of high strength SMA weld metal'. Abstracts, 76th AWS Annual Convention, Cleveland, 1995.

5

Effects of heat treatment

Two aspects of heat treatment were examined,[1] stress relief (or post-weld heat treatment, PWHT) and normalising and tempering (N + T).[2] Both treatments were examined over a range of manganese contents, stress relief being carried out on deposits with a range of carbon and manganese contents, whereas the normalising experiments were carried out with varying manganese only. Like manganese itself, most other elements examined subsequently were tested in both as-welded and stress relieved conditions and the results are to be found in the appropriate chapter.

Stress relief

Deposits of the ISO 2560 type (Ch. 2, Ref. 3) with three beads/layer were made in the usual way, with an interpass temperature of 200 °C; test specimens were heat treated for 2 hours at 580 °C, except for some specimens of one series (1.8%Mn, 0.15%C), which were given extended heat treatments for times up to 100 hours. The compositions of the deposits were similar to those for the as-welded series in Table 3.6 and are not reproduced here.

Metallography

Because stress relief heat treatment has an imperceptible effect on microstructures in the light microscope, examination was restricted to electron microscopy (at TWI) of the high and low carbon deposits from 1.4%Mn electrodes (code C). Specimens from the top run and the underlying coarse- and fine-grained reheated regions were selected, electropolished, etched in 2% nital and carbon replicas taken. Comparison was made with the as-welded deposits of similar composition examined in Chapter 3.

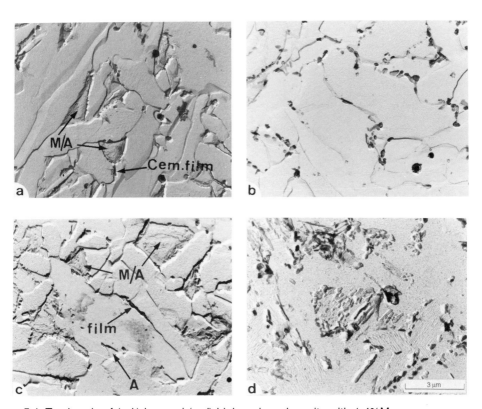

5.1 Top beads of (*a,b*) low and (*c,d*) high carbon deposits with 1.4%Mn: *a,c*) As-welded; *b,d*) Stress-relieved at 580 °C (A – retained austenite, B/P – bainite/pearlite, Cem – cementite, M/A – martensite/austenite, P – pearlite).

The top, as welded, beads of both deposits (Fig. 5.1(*a,c*)), contained acicular ferrite with retained austenite partially transformed to martensite (M/A constituent, more of which was present with the higher carbon content) and a little cementite film. Stress relief of low carbon deposits led to discrete grain boundary carbides and, with high carbon weld metal, to carbide aggregates (Fig. 5.1(*b,d*)).

The coarse-grained reheated regions (Fig. 5.2(*a,c*)) contained ferrite with cementite films and pearlite or, with higher carbon, bainite/pearlite and M/A constituent. Stress relieving (Fig. 5.2(*b,d*)) partially broke down the pearlite and spheroidised the films whilst the M/A constituent was replaced by carbide aggregates.

The fine-grained regions contained retained austenite, partially transformed to martensite, with some cementite films and

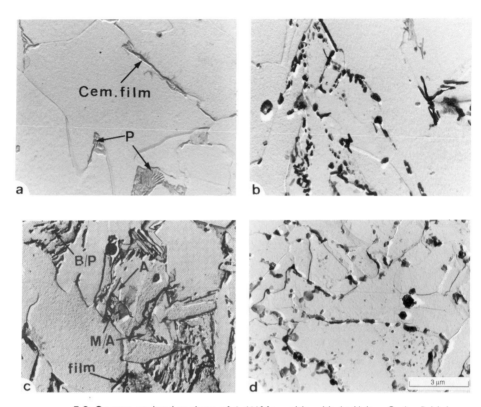

5.2 Coarse-grained regions of 1.4%Mn welds with (a,b) low C, (c,d) high C: a,c) As-welded; b,d) Stress relieved at 580 °C (key as Fig. 5.1).

bainite/pearlite (Fig. 5.3(a,c)). The higher carbon deposit contained more pearlite and more retained austenite, which had transformed to martensite, than the lower carbon deposit. Stress relief (Fig. 5.3(b,d)) broke down and spheroidised the pearlite and precipitated grain boundary carbides and transformed the retained austenite.

Mechanical properties

Hardness levels were reduced by stress relieving – particularly in the case of top (untempered) beads. The variation in hardness on traversing the centre of the weld from the top bead was also decreased (Fig. 5.4). The influence of carbon content on top bead hardness levels is shown in Fig. 5.5, which may be compared with Fig. 3.24 for the corresponding as-welded curves.

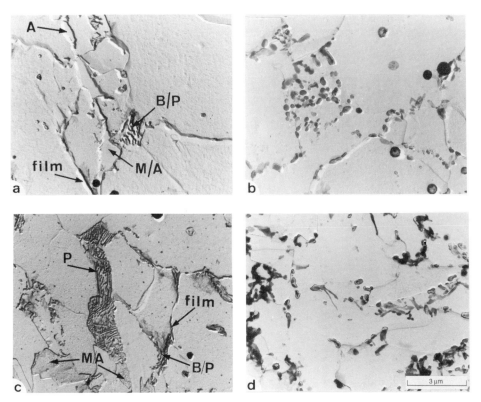

5.3 Fine-grained regions of 1.4%Mn welds with (a,b) low C, (c,d) high C: a,c) As-welded; b,d) Stress-relieved at 580 °C (key as Fig. 5.1).

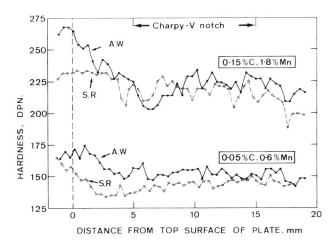

5.4 Comparison of hardness traverses (HV5) down the centreline of welds deposited with electrodes A and C, as-welded (AW) and stress-relieved at 580 °C (SR).

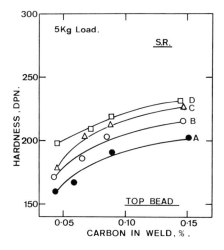

5.5 Influence of carbon on top bead hardness of deposits of varying manganese level stress-relieved at 580 °C.

Table 5.1 Tensile test results from welds of varying carbon and manganese content, stress-relieved at 580°C

C level, %	Electrode code	Yield strength, N/mm^2	Tensile strength, N/mm^2	Elongation, %	Reduction of area, %
0.044	A	370	450	34	83
	B	390	480	34	81
	C	410	500	31	81
	D	440	520	30	80
0.064	A	390	460	33	79
	B	420	500	35	81
	C	460	540	31	78
	D	500	590	26	75
0.096	A	390	500	34	79
	B	440	550	34	77
	C	460	570	29	78
	D	490	600	30	75
0.147	A	450	530	31	76
	B	480	580	29	77
	C	510	620	27	73
	D	570	680	24	70

Note: for comparable as-welded results see Table 3.7.

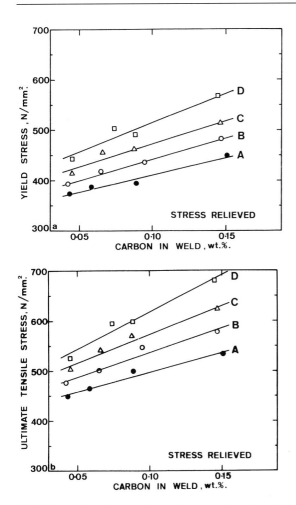

5.6 Effect of carbon on strength properties of multirun deposits with varying manganese content stress-relieved at 580 °C: *a*) Yield strength; *b*) Tensile strength.

Stress-relief reduced yield strength by approximately 33 N/mm^2 and tensile strength by ~12 N/mm^2 (Table 5.1), thus reducing the yield/tensile ratio to values within the range 0.78–0.85, compared with 0.84–0.90 in the as-welded condition. Linear relationships between strength levels and carbon (Fig. 5.6) led to the following equations:

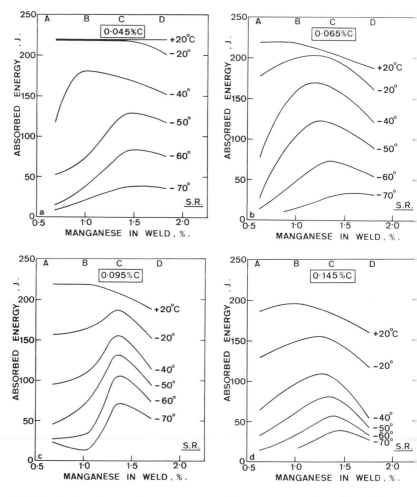

5.7 Effect of carbon and manganese on Charpy energy values for welds stress-relieved at 580 °C.

$$YS_{sr} = 310 + 391\ C + 50\ Mn + 429\ \left(C.Mn\right) \qquad [5.1]$$

$$TS_{sr} = 396 + 330\ C + 42\ Mn + 643\ \left(C.Mn\right) \qquad [5.2]$$

The equations show lower factors for C and Mn than the corresponding equations for as-welded strength (Eq. [3.5] and [3.6]), although the factor for C.Mn was higher.

Plots of the dependence of Charpy energy on manganese content (Fig. 5.7), show an optimum manganese content close to 1.4%, as in the as-welded condition. Figure 5.8 shows how the shift in Charpy transition temperature on stress relief varied with manganese content.

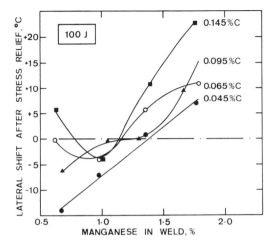

5.8 Lateral shift in 100 J Charpy transition temperature of welds with varying carbon and manganese content stress-relieved at 580 °C.

Stress relief was beneficial at low manganese levels (particularly at the lowest carbon content) and was detrimental at high manganese (particularly at the highest carbon content). At intermediate manganese levels, stress relieving had little influence on the Charpy transition temperatures. Otherwise, the influence of carbon on toughness was similar to the as-welded condition, i.e. carbon decreased upper shelf values and made the transition between brittle and ductile fracture more gradual. However, these relatively small changes in Charpy values on stress relief would, in a practical situation, be relatively small compared with the beneficial effect of reducing residual stresses by PWHT.

The results of additional tests of up to 100 hours duration, carried out on the weld showing the greatest drop of toughness on PWHT (0.145%C, 1.8%Mn) are illustrated in Fig. 5.9. Charpy toughness, in terms of 100 and 28 J transition temperatures, deteriorated on increasing the time of PWHT from 1 to 20 hours and then improved. Electron micrographs (Fig. 5.10) showed a coarsening and rounding of the precipitated carbides during this period.

Discussion

During stress relief three mechanisms are likely. The first, giving an improvement in toughness, is softening of ferrite and tempering of any martensite in the M/A constituent. Grain boundary carbides, particularly in the fine-grained HAZ regions, will change into discrete

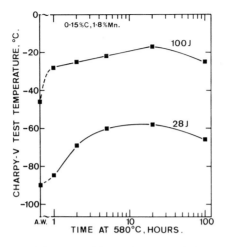

5.9 Influence of time of stress relief on 100 J and 28 J Charpy transition temperature of 0.145%C, 1.8%Mn weld.

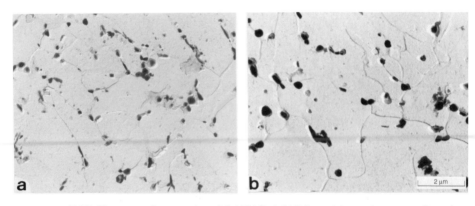

5.10 Electron micrographs of 0.15%C–1.8%Mn weld metal stress-relieved at 580 °C for: *a*) 2 hours; *b*) 100 hours.

carbides. These beneficial mechanisms are opposed by the formation of relatively brittle carbides from residual austenite and coarsening of carbides in pearlite and elsewhere. The welds of highest manganese content, containing the most prior austenite, will have given the greatest precipitation of carbides and the greatest (albeit small) reduction in toughness. Finally, a decrease in dislocation density will contribute to a reduction in strength and, hence, an improvement in toughness.

Normalising and tempering

Portions of the welds of varying manganese content, whose compositions are given in Table 3.1, were selected for the normalising and tempering experiments.[2] The standard treatment was 30 min heating at 930 °C, either without tempering (N) or after tempering for 2 hours at 580 °C (N + T). Additional tests on the 1.4%Mn weld (electrode C) were made with normalising temperature varying from 880 to 1080 °C for 30 min, again with and without tempering for 2 hours at 580 °C.

Metallography

After normalising, the microstructure consisted of ferrite with small amounts of other constituents (Fig. 5.11), the ferrite being appreciably coarser than the fine-grained regions of as-welded deposits, e.g. Fig. 3.5 and 3.8. Manganese refined the ferrite grain size as shown in Fig. 5.12, the effect becoming marked above 1.4%Mn.

At low manganese levels, cementite films, continuous or discontinuous at the ferrite boundaries (Fig. 5.13) and small pearlite colonies were present. Increasing manganese levels reduced the amounts of cementite films but introduced a complex constituent (designated M/A/B) consisting of martensite, retained austenite and bainite, which is shown in Fig. 5.14. The influence of manganese on the proportions of the additional constituents, measured (by BSC) in an SEM at ×5000, is shown in Fig. 5.15.

Tempering at 580 °C globularised the cementite films, partially broke down the pearlite and dissociated the M/A/B constituent, leaving fine carbide particles (Fig. 5.16).

Heating at 880 °C gave only partial transformation to austenite (Fig. 5.17). Normalising at higher temperatures coarsened the grain size (Table 5.2), although not excessively, and gave larger and better defined pearlite colonies (Fig. 5.18). Grain size measurements on normalised and tempered deposits gave virtually identical results.

Mechanical properties

Hardness traverses down the centreline of normalised welds (Fig. 5.19) showed relatively low levels of hardness with little variation compared with the as-welded or stress-relieved conditions (Fig. 5.4). Hardness increased from ~115 HV5 with 0.6%Mn to ~145 HV5 with 1.8%Mn.

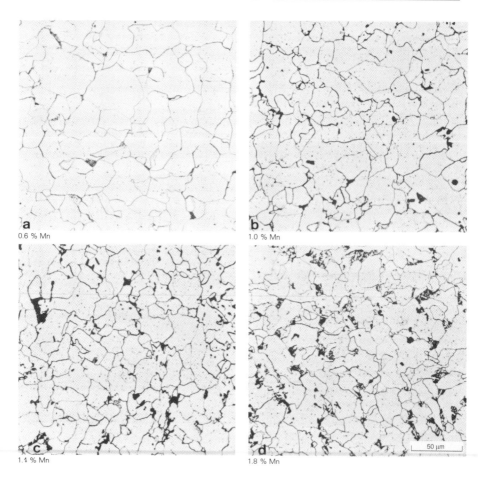

5.11 Microstructure of welds normalised at 930 °C with: *a*) 0.6%; *b*) 1.0%; *c*) 1.4%; *d*) 1.8%Mn.

Tensile properties in normalised and in normalised and tempered conditions, given in Table 5.3, show that yield strength values were considerably lower than for the as-welded or stress-relieved conditions (Table 3.5), generally by over 100 N/mm², although tensile strength and ductility values were considerably less affected. Strength values after tempering following the normalising were mostly no more than 10 N/mm² different from the normalised values. The equations for the best fit straight lines relating strength to manganese content were:

$$YS_N = 285 + 20 \ Mn \hspace{4em} [5.3]$$
$$TS_N = 353 + 80 \ Mn \hspace{4em} [5.4]$$

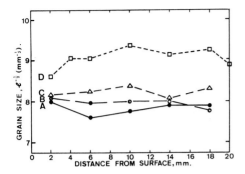

5.12 Ferrite grain size variation down the centreline of deposits of varying manganese content normalised at 930 °C.

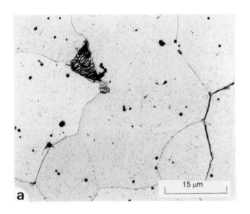

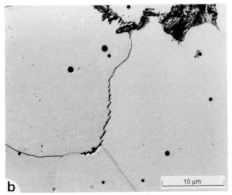

5.13 Microstructure of welds normalised at 930 °C: *a*) With 0.6%Mn showing continuous cementite film and pearlite colony; *b*) With 1.0%Mn showing discontinuous film and pearlite.

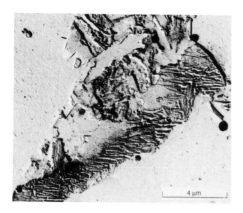

5.14 Microstructure of 1.8%Mn weld normalised at 930 °C showing M/A/B constituent, TEM replica.

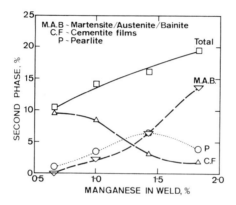

5.15 Influence of manganese on volume fraction of microstructural constituents of welds normalised at 930 °C.

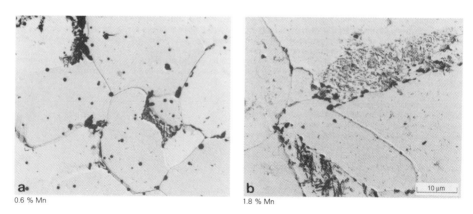

5.16 Microstructure of welds normalised at 930 °C and tempered at 580 °C: *a*) 0.6%Mn; *b*) 1.8%Mn.

Table 5.2 Mean linear intercept ferrite grain size of 1.4%Mn weld normalised at different temperature

Normalising temperature, °C	Mean intercept ferrite grain size, mm$^{-1/2}$	
	Normalised	Normalised and tempered
880	9.7	9.5
930	8.6	8.6
980	8.5	8.5
1030	7.8	7.5
1080	7.2	7.3

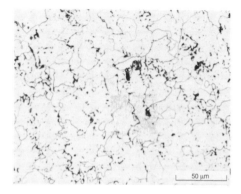

5.17 Microstructure of 1.4%Mn weld heat treated at 880 °C showing partial normalising.

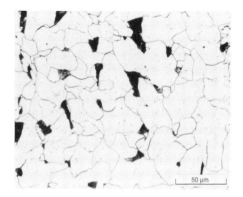

5.18 Microstructure of 1.4%Mn weld normalised at 1030 °C.

Table 5.3 Tensile properties of normalised deposits with 1.4%Mn

Heat treatment	Electrode code	Yield strength, N/mm²	Tensile strength, N/mm²	Elongation, %	Reduction of area, %
N	A	300	410	37	81
	B	300	430	38	81
	C	320	460	35	80
	D	320	510	34	80
N + T	A	300	400	40	81
	B	300	420	37	82
	C	310	450	38	80
	D	340	480	35	80

Notes: N – normalised 0.5 hour at 930 °C.
N + T – normalised as above and tempered 2 hour at 580 °C.
Corresponding as-welded and stress-relieved results are in Table 3.5.

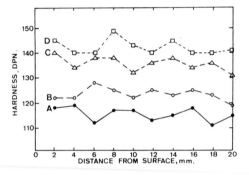

5.19 Hardness traverses (HV5) down the centreline of deposits of varying manganese content normalised at 930 °C.

$$YS_{N+T} = 269 + 35\ Mn \qquad [5.5]$$

$$TS_{N+T} = 362 + 62\ Mn \qquad [5.6]$$

Compared with the as-welded and stress-relieved equations (Eq. [3.1]–[3.4]), manganese has a considerably reduced effect on both yield strength and, to a lesser degree, on tensile strength, as can be seen from Fig. 5.20.

The main reason for the considerable reduction in yield strength is the destruction, by re-austenitising and slow cooling, of the fine dislocation network present in as-welded deposits, which persists to a considerable extent after stress relief (Ch. 1, Ref. 1). The coarser

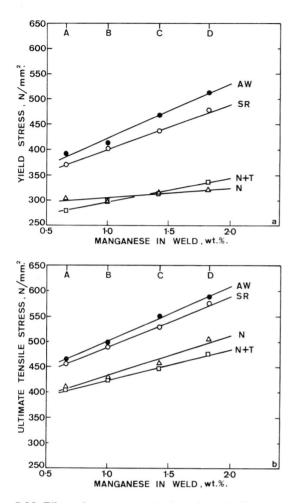

5.20 Effect of manganese in deposits with different heat treatments on:
a) Yield strength; b) Tensile strength (AW – as-welded, N – normalised at
930 °C, N + T – normalised at 930 °C and tempered at 580 °C, SR –
stress-relieved at 580 °C).

ferrite grain size of normalised weld metal compared with the generally finer as-welded microstructure must also play a part.

The temperature of normalising of the 1.4%Mn deposit (Table 5.4)
had little effect on the levels of strength or ductility; subsequent tempering at 580 °C slightly reduced strength levels, regardless of the normalising temperature.

Table 5.4 Effect of normalising temperature on tensile properties of 1.4%Mn weld metal

Heat treatment	Electrode code	Yield strength, N/mm^2	Tensile strength, N/mm^2	Elongation, %	Reduction of area, %
N 880°C		320	470	37	82
N 930°C		320	460	35	82
N 980°C		330	470	39	81
N 1030°C		330	460	38	81
N 1080°C		350	460	37	81
N 880°C + T		300	440	36	82
N 930°C + T		310	450	38	80
N 980°C + T		300	440	37	82
N 1030°C + T		300	440	38	81
N 1080°C + T		310	440	36	81

Note: N – normalised for 0.5 hour at stated temperature; T – tempered 580°C/2 hour.

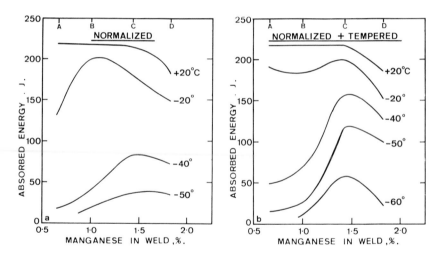

5.21 Influence of manganese content on Charpy energy at varying temperature for welds: *a*) Normalised at 930 °C; *b*) Normalised and tempered at 580 °C.

From the Charpy transition curves, plots showing the influence of manganese on the derived parameters were prepared (Fig. 5.21 and 5.22). Manganese improved toughness and the optimum level appeared to be about 1.4% for both the normalised and the normalised and tempered conditions. Tempering improved Charpy

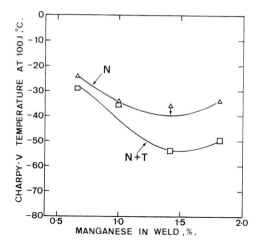

5.22 Influence of manganese on 100 J Charpy transition temperature for welds normalised at 930 °C and normalised and tempered at 580 °C.

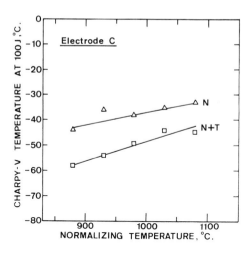

5.23 Influence of normalising temperature on 100 J Charpy transition temperature of deposits with 1.4%Mn normalised and normalised and tempered at 580 °C.

toughness, particularly at the two higher manganese levels, presumably as a result of breaking up the cementite films at the lowest manganese level and tempering of the martensite in the complex M/A/B constituent at the highest level.

Nevertheless, the relatively good Charpy levels must be seen in the context of the low yield strength levels of the normalised weld metal. Although the normalised 1.4%Mn weld metal gave similar Charpy values over a range of temperature to the as-welded or stress-relieved values, the normalised yield strength was over 150 N/mm^2 lower.

Increasing the normalising temperature led to a deterioration in toughness, as shown in Fig. 5.23 for the 1.4%Mn weld, although this was partly improved by subsequent tempering. These trends were maintained down to a normalising temperature of 880 °C, even though normalising was incomplete at this temperature.

References

1 Evans GM: 'Effect of stress relieving on the microstructure and properties of C–Mn all-weld-metal deposits'. *Welding J* 1986 **65**(12) 326s–34s; *Welding Res Abroad* 1991 **37**(2/3) 2–10; IIW Doc II–A–583–83; *Schweiss-mitteilungen* 1983 **41**(103) 15–27.
2 Evans GM: 'Effect of heat treatment on the microstructure and properties of C–Mn all-weld-metal deposits'. *Metal Construction* 1985 **17**(10) 676R–82R; *Welding Res Abroad* 1991 **37**(2/3) 11–19; IIW Doc II–A–583–83; IIW Doc II–A–605–84; *Schweissmitteilungen* 1985 **43**(107) 16–27.

6

Silicon and impurities

Although sometimes regarded as impurities, the elements silicon, sulphur and phosphorus can be deliberately added to some weld metals and steels. Silicon is added to most weld metals as a deoxidant; indeed, its absence from MMA and TIG welds can lead to porosity. Sulphur is a naturally occurring impurity in steels and in some flux minerals but is deliberately added to steels of the free-cutting type, from which it can be diluted into weld metals. With basic covered electrodes, such as those used for the present work, some (but by no means all) of the sulphur in the weld metal is removed, principally by calcium fluoride in the flux coating. Phosphorus is another natural impurity in steels and some flux minerals and, at relatively high heat inputs, is known to reduce toughness in weld metal. Finally, arsenic is typical of the impurities not normally analysed for, except in Cr–Mo steels, where resistance to temper embrittlement is required and low levels of As, Sb and Sn are essential.

Electrodes were made to deposit welds having the expected range of these elements in weld metal made with good and poor quality constituents and tested both as-welded and, except in the case of arsenic, after stress relief of 2 hours at 580 °C.

Silicon (Mn–Si)

Test welds were deposited with electrodes yielding the four standard manganese levels (A–D), with silicon contents ranging from 0.2 to 0.9% and tested as-welded and after stress relief.[1] The mean compositions of the as-welded and stress relieved welds are given in Table 6.1.

Apart from the elements varied intentionally, the only element to be affected by silicon was oxygen, which decreased as the silicon content was increased, as would be expected by increasing the amount of a deoxidant in the weld metal. Although it was possible to balance

Table 6.1 Composition of weld metal with varying silicon and manganese content

Element, wt%		C	Mn	Si	N	O	Mn/Si ratio
Si level, %	Electrode code						
0.2	A	0.060	0.60	0.20	0.006	0.050	3.0
	B	0.063	0.99	0.20	0.006	0.046	5.0
	C	0.064	1.40	0.20	0.006	0.044	7.0
	D	0.064	1.82	0.19	0.007	0.044	9.6
0.4	A	0.070	0.66	0.38	nd	0.045	1.7
	B	0.065	1.04	0.38	nd	0.042	2.7
	C	0.066	1.41	0.36	nd	0.042	3.7
	D	0.067	1.80	0.35	nd	0.041	5.1
0.6	A	0.065	0.65	0.61	nd	0.040	1.1
	B	0.062	1.03	0.63	nd	0.039	1.6
	C	0.073	1.44	0.62	nd	0.037	2.3
	D	0.068	1.78	0.59	nd	0.038	3.0
0.9	A	0.070	0.64	0.95	0.006	0.035	0.7
	B	0.065	0.99	0.95	0.005	0.035	1.0
	C	0.063	1.38	0.93	0.005	0.034	1.5
	D	0.064	1.75	0.92	0.006	0.030	1.9

Notes: all welds contained 0.005–0.007%S and 0.007–0.009%P; nd – not determined. Typical levels of other elements are given in Table 3.1.

other elements as silicon and manganese were altered, it would not have been possible to balance the oxygen content without adding another deoxidant, such as magnesium or titanium, which would have been outside the purpose of these tests. Plotting weld oxygen against weld silicon (Fig. 6.1) showed a wide scatter band with the lower manganese welds lying at the top and the higher manganese welds at the bottom, consistent with manganese also acting as a deoxidant, albeit less powerfully. Regression of the data gave a linear equation, plotted in Fig. 6.2, with little scatter:

$$[O]_{(ppm)} = 541 - 165 \text{ Si} - 43.5 \text{ Mn} \qquad [6.1]$$

$$[O]_{(ppm)} = 541 - 165 \left(\text{Si} + \text{Mn}/3.8\right) \qquad [6.1a]$$

Plotting weld oxygen against the Mn/Si ratio (Fig. 6.3), shows that silicon and manganese exerted individual effects. The value of Mn/3.8 in the alternative form of the above equation (Eq. [6.1a]), suggests that this value is related to the difference in atomic weights (Mn = 55, Si = 28) and that Si is being quadrivalent (SiO_2) and Mn divalent (MnO), thus giving a theoretical ratio of 55/14 = 3.9, close to the value of 3.8 found.

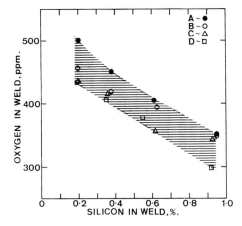

6.1 Influence of silicon on weld oxygen content.

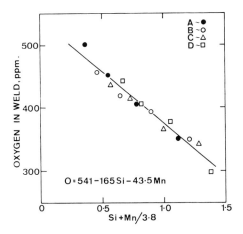

6.2 Combined influence of silicon and manganese on weld oxygen content.

Metallography

At high levels of manganese, silicon caused no outstanding changes to the microstructure of the as-welded, columnar region, although point counting revealed significant changes at lower manganese contents (Fig. 6.4). With 0.6%Mn, silicon increased the proportion of acicular ferrite, albeit somewhat irregularly, at the expense of other constituents. With higher manganese contents, silicon increased the aspect ratio of the acicular ferrite, in line with other observations on

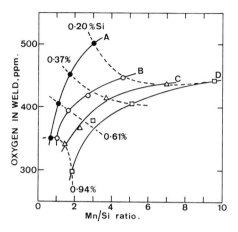

6.3 Mn/Si ratio against weld oxygen content.

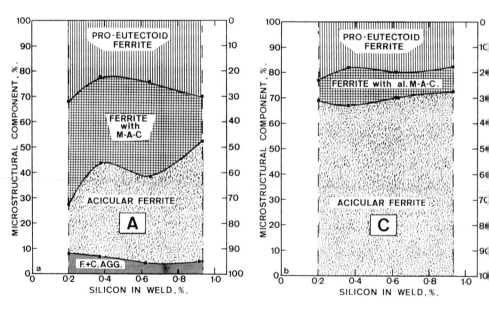

6.4 Influence of silicon on: a) 0.6%Mn; b) 1.4%Mn as-deposited microstructure.

submerged-arc weld metal[2] although, unlike the submerged-arc deposits, no refinement of the acicular ferrite was noted. Silicon also increased the proportion of dark-etching constituent between the ferrite laths (Fig. 6.5). Examination of extraction replicas (Fig. 6.6)

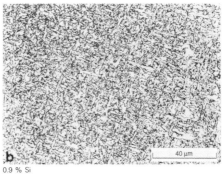

0.2 % Si   0.9 % Si

6.5 Acicular ferrite microstructure in as-deposited weld metal with 1.4%Mn and: *a*) 0.2%Si; *b*) 0.9%Si.

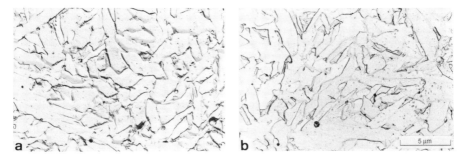

6.6 Carbon replicas of acicular ferrite in as-deposited weld metal with 1.4%Mn and: *a*) 0.2%Si; *b*) 0.9%Si.

showed the dark-etching regions to be retained austenite or the M/A constituent. Stress relief led to nucleation and growth of carbides in these regions.

Little effect of silicon was apparent in the coarse-grained reheated regions. The grain size of the fine-grained reheated regions was not affected by silicon, averaging 5.0 μm for the welds made with electrode C. However, silicon increased the proportion of constituents other than ferrite in this region, as shown for the 1.4%Mn weld metal in Fig. 6.7(*a*). The main increase was in the M/A constituent at the expense of cementite films (Fig. 6.8). In contrast, at the 0.4%Si level, manganese gave a smaller increase in M/A, coupled with a rise in bainite/pearlite (B/P) (Fig. 6.7(*b*)).

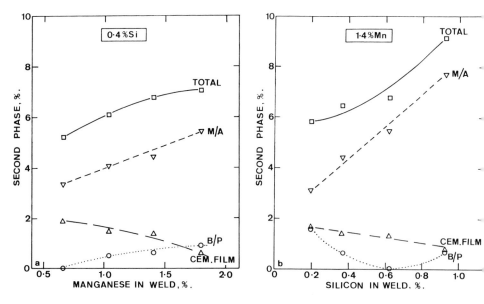

6.7 Influence on constitution of the fine reheated microstructure of: *a*) Manganese with 0.4%Si; *b*) Silicon with 1.4%Mn.

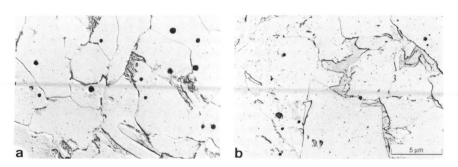

6.8 Carbon replicas of ferrite in fine reheated 1.4%Mn weld metal with: *a*) 0.2%Si showing degenerate pearlite (B/P constituent); *b*) 0.9%Si with M/A constituent.

Mechanical properties

Hardness surveys in the as-welded condition showed silicon to increase hardness at a decreasing rate as more silicon was added (Fig. 6.9). Regression analysis of the data gave the following best fit equation:

$$HV5 = 107 + 56 \text{ Mn} + 158 \text{ Si} - 57 \text{ Si}^2 - 39 \text{ Mn. Si} \qquad [6.2]$$

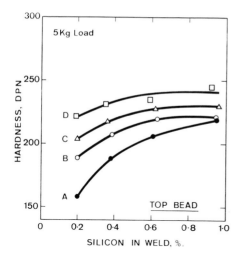

6.9 Influence of silicon on as-welded hardness of welds with varying manganese content.

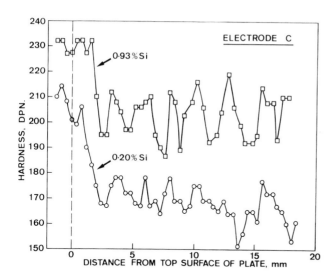

6.10 As-welded hardness surveys down the centreline of welds with 1.4%Mn and varying silicon content.

Centreline hardness traverses (Fig. 6.10) on welds with 1.4%Mn showed the top, un-reheated bead to be about 30 HV5 harder than the underlying weld metal. The highest silicon deposit showed more variation in hardness than the 0.2%Si deposit.

Tensile properties are given in Table 6.2. and plotted in Fig. 6.11.

Table 6.2 Tensile test results on welds of varying silicon and manganese content

Si level, %	Condition	Electrode code	Yield strength, N/mm²	Tensile strength, N/mm²	Elongation, %	RA, %
0.2	AW	A	390	450	34	80
		B	420	480	33	80
		C	450	510	32	81
		D	500	550	29	80
0.4	AW	A	420	490	34	80
		B	460	520	30	81
		C	490	550	31	79
		D	530	590	29	79
0.6	AW	A	440	500	33	79
		B	470	530	30	80
		C	520	580	29	77
		D	560	610	29	76
0.9	AW	A	480	550	28	73
		B	480	570	29	77
		C	520	600	29	78
		D	560	640	29	77
0.2	SR	A	360	440	37	80
		B	400	470	35	81
		C	420	500	31	79
		D	470	540	29	79
0.4	SR	A	390	470	35	81
		B	430	500	32	81
		C	450	540	32	79
		D	490	560	29	77
0.6	SR	A	400	500	33	73
		B	420	520	33	80
		C	460	570	31	76
		D	510	610	28	75
0.9	SR	A	440	540	30	75
		B	450	550	31	77
		C	500	600	30	76
		D	540	630	29	76

Note: RA – reduction of area; AW – as-welded; SR – stress relieved.

As with hardness, silicon increased strength properties less as more was added. Regression analysis showed that this could be accounted for by a negative Si^2 term:

$$YS_{aw} = 293 + 91\ Mn + 228\ Si - 122\ Si^2 \qquad [6.3]$$

$$TS_{aw} = 365 + 89\ Mn + 169\ Si - 44\ Si^2 \qquad [6.4]$$

$$YS_{sr} = 288 + 91\ Mn + 95\ Si - 10\ Si^2 \qquad [6.5]$$

$$TS_{sr} = 344 + 89\ Mn + 212\ Si - 79\ Si^2 \qquad [6.6]$$

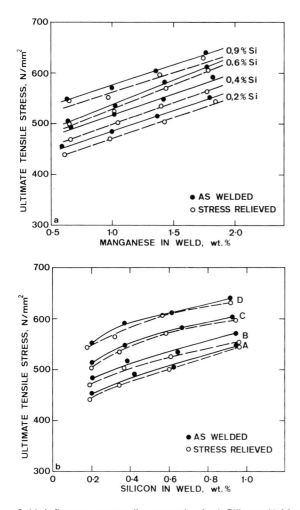

6.11 Influence on tensile strength of: *a*) Silicon; *b*) Manganese.

The factors for manganese are not very different from those in Eq. [3.1]–[3.4] and, although the factors for silicon are mostly greater than those for manganese, the effect of silicon at the higher silicon level is similar to or less than that of manganese.

Silicon reduced ductility somewhat, the effect on elongation being greater than on the reduction of area. Strength was reduced by stress relief, the reduction in yield strength being greater than for tensile strength.

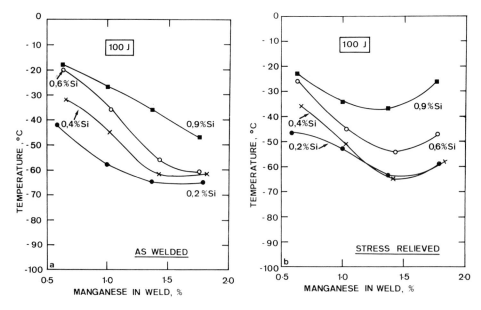

6.12 Influence of silicon and manganese on 100 J Charpy transition temperature: *a*) As-welded; *b*) After stress relief at 580 °C.

Charpy toughness deteriorated as silicon was increased (Fig. 6.12). With manganese up to 1%, the deterioration was most marked at low silicon levels. However, at higher manganese contents, up to about 0.5%Si was acceptable – fortunately in view of its beneficial effects on welding behaviour. Figure 6.12 shows that the hitherto optimum manganese content of about 1.4% is clearly demonstrated in the stress-relieved condition but tends to a higher manganese content as-welded, particularly at silicon levels above 0.4%.

The pattern of Charpy behaviour after stress relief was similar to that with varying carbon and manganese shown in Fig. 5.8, i.e. stress relief was mildly beneficial at low manganese levels but harmful when the manganese content was high, particularly at high silicon levels.

Sulphur

Welds were made with 1.4%Mn electrodes (coded C) with added iron sulphide to give weld sulphur contents ranging from 0.007 to 0.046%.[3] Tests were carried out as-welded and after 2 hours stress relief at 580 °C. Weld compositions are given in Table 6.3. Apart from

Table 6.3 Composition of 1.4%Mn weld metal having varying sulphur content

Element, wt%						
C	Mn	Si	S	P	N	O
0.063	1.39	0.37	0.007	0.007	0.005	0.044
0.057	1.41	0.37	0.016	0.007	0.006	0.046
0.057	1.38	0.35	0.027	0.007	0.006	0.048
0.062	1.39	0.36	0.038	0.007	0.005	0.050
0.063	1.38	0.35	0.046	0.008	0.005	0.052

Note: typical levels of other elements are given in Table 3.1.

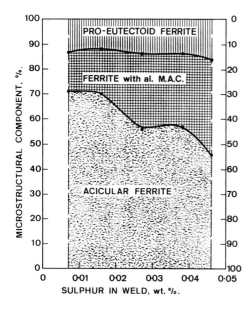

6.13 Influence of sulphur on microstructural constituents of as-deposited 1.4%Mn weld metal ('al MAC' is now known as ferrite with aligned second phase (FS–A)).

sulphur, the only element to vary was oxygen, which increased slightly as sulphur was increased.

Metallography

In the top, un-reheated weld beads, increasing the sulphur content decreased the proportion of acicular ferrite, which was replaced by

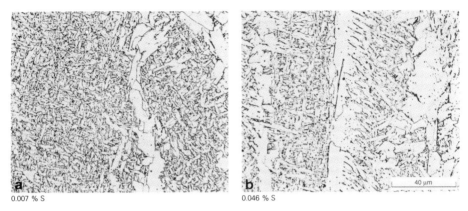

0.007 % S

0.046 % S

6.14 Microstructure of as-deposited (top run) regions of 1.4%Mn welds with: *a*) 0.007%S; *b*) 0.046%S.

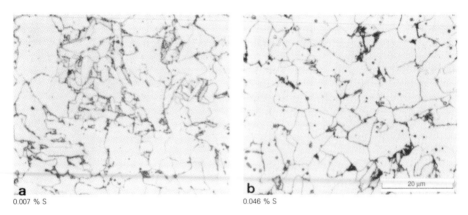

0.007 % S

0.046 % S

6.15 Microstructure of fine-grained regions of 1.4%Mn deposits with: *a*) 0.007%S; *b*) 0.046%S.

ferrite with aligned second phase (Fig. 6.13 and 6.14). Little change was apparent in reheated weld metal except that, in the fine-grained region, increasing sulphur changed the structure (Fig. 6.15) from ferrite with light-etching M/A to the darker degenerate pearlite (B/P).

The inclusion content was considerably increased by increasing the weld sulphur content. Using the estimation technique developed by Widgery,[4] the increase in sulphur was calculated nearly to double the weld inclusion volume fraction (from 0.32 to 0.57% by volume), largely as a result of the increasing number of manganese sulphide

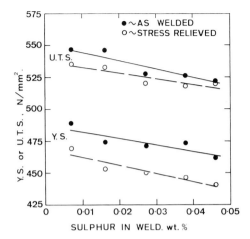

6.16 Influence of sulphur on strength properties of 1.4%Mn weld metal.

(MnS) inclusions; all microsections examined showed a considerable increase, although no measurements were made. Details of investigations into the composition and structure of typical inclusions are given in Chapter 17.

Mechanical properties

Hardness values were only slightly affected by sulphur, the reduction of no more than 10 HV5 being consistent with the removal of Mn from the metallic matrix to form MnS. Increasing sulphur also reduced the variation in hardness on traversing down the weld centreline. Because of the relatively large number of inclusions, even in the low sulphur weld metal, there was no evidence that hardness was being reduced by a mechanism seen in parent steels and HAZs,[5] where an increase in sulphur reduces the hardenability of a HAZ as a result of inclusions providing sites at which ferrite can easily nucleate at high temperatures.

The tensile properties in Table 6.4 show slight falls in both strength (Fig. 6.16) and reduction of area as the weld sulphur content was increased. Elongation values were unaffected, presumably because they were less influenced by the inclusions present than was the reduction of area. Regression analysis showed the extent of the drop in yield strength and tensile strength:

$$YS_{aw} = 489 - 596\ S \qquad\qquad [6.7]$$
$$TS_{aw} = 547 - 524\ S \qquad\qquad [6.8]$$

Table 6.4 Tensile properties of 1.4%Mn welds of varying sulphur content

Condition	S, wt%	Yield strength, N/mm²	Tensile strength, N/mm²	Elongation, %	Reduction of area, %
AW	0.007	490	550	30	79
	0.016	470	550	31	77
	0.027	470	530	31	75
	0.038	470	530	31	73
	0.046	460	520	31	74
SR	0.007	470	540	29	78
	0.016	450	530	31	77
	0.027	450	520	30	75
	0.038	450	520	30	75
	0.046	440	520	30	72

Note: AW – as-welded; SR – stress relieved.

$$YS_{sr} = 471 - 757 \, S \qquad [6.9]$$
$$TS_{sr} = 544 - 406 \, S \qquad [6.10]$$

Over the full range of sulphur levels, the fall in strength was about 20 N/mm², rather more than could be accounted for by the removal of manganese to form MnS inclusions.

Charpy tests showed a consistent rise in transition temperature, a fall in upper shelf energy and a decrease in the slope of the transition curves as sulphur was increased (Fig. 6.17). The Charpy temperatures for 100 and 28 J are plotted in Fig. 6.18 and illustrates the extent of embrittlement. The present results agree well with earlier results on MMA welds reported by Steel,[6] showing the fall in upper shelf toughness with sulphur, although Steel's results, in which low sulphur weld metal contained less acicular ferrite than in the present deposits, did not show a significant change in transition temperature. The reason for this difference is not known, other than a speculation that, because Steel's base composition contained less acicular ferrite than the present welds, it lost less on adding sulphur, so that his increase in transition temperature would have been negligible.

Other work on as-welded submerged-arc welds (Ref. 8, Ch. 3) developed linear relationships between the estimated inclusion content and Charpy upper shelf energy and also the slope of the transition curves. The upper shelf energy in Ref. 8, Ch. 3 fell by ~8 J for each 0.1% (by volume) of inclusions. This would give a much smaller decrease in upper shelf energy (~21 J) than was actually found in the current work, i.e. nearly 100 J over the full range of sulphur content.

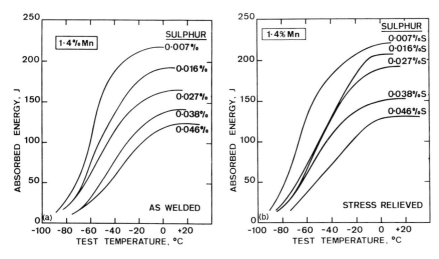

6.17 Influence of sulphur on Charpy V transition curves of 1.4%Mn weld metal: *a*) As-welded; *b*) Stress-relieved at 580 °C.

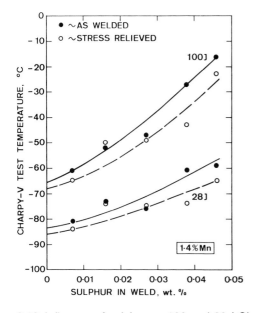

6.18 Influence of sulphur on 100 and 28 J Charpy temperature.

Table 6.5 Composition of 1.4%Mn weld metal at three heat inputs having varying phosphorus content

Heat input, kJ/mm	Element, wt%				
	C	Mn	Si	S	P
1.0	0.067	1.43	0.38	0.005	0.007
	0.069	1.37	0.39	0.005	0.015
	0.066	1.40	0.37	0.005	0.023
	0.066	1.36	0.35	0.005	0.030
	0.065	1.45	0.35	0.006	0.040
2.2	0.059	1.23	0.35	0.007	0.008
	0.055	1.24	0.35	0.008	0.017
	0.053	1.20	0.32	0.008	0.024
	0.055	1.17	0.32	0.009	0.033
	0.053	1.17	0.31	0.009	0.040
4.3	0.057	1.20	0.32	0.006	0.009
	0.055	1.19	0.30	0.007	0.018
	0.050	1.13	0.28	0.007	0.029
	0.051	1.09	0.27	0.007	0.041
	0.052	1.13	0.27	0.007	0.049

Notes: the welds at 1 kJ/mm contained 0.006%N and 0.042–0.043%O.
Typical levels of other elements are given in Table 3.1.

Phosphorus

Two series of welds were made, the first with 1.4%Mn electrodes used at a heat input of 1 kJ/mm as in the bulk of the work.[3] Because of concern about the effect of heat input on possible embrittlement by phosphorus, further welds[7] were made at heat inputs of 2.2 and 4.3 kJ/mm; these were achieved as in Chapter 4, in the section *Heat input*. In all cases the electrode coatings were modified with ferrous phosphate (Fig. 2.3(*f*)) to give weld metal phosphorus contents between 0.007 and 0.040%. Welds were tested as-welded and, in the case of the 1 kJ/mm deposits only, after 2 hours stress relief at 580 °C.

Weld compositions, given in Table 6.5, were satisfactorily uniform within each series. As in the results reported in Chapter 4 (Table 4.10), increasing heat input tended to reduce the contents of C, Mn and Si. At the highest heat input, the content of phosphorus was noticeably greater, particularly at the higher levels.

Metallography

There were no significant changes in microstructure visible in the light microscope. At 1 kJ/mm, top run deposits contained between

Table 6.6 Tensile properties of 1.4%Mn welds of varying phosphorus content

Condition	P, wt%	Yield strength, N/mm²	Tensile strength, N/mm²	Elongation, %	Reduction of area, %
1 kJ/mm, AW	0.007	510	560	31	78
	0.015	520	570	31	76
	0.023	500	560	30	79
	0.030	510	560	29	78
	0.040	510	570	29	78
2.2 kJ/mm, AW	0.008	420	500	31	81
	0.007	430	490	24	81
	0.024	430	510	30	81
	0.033	420	490	31	81
	0.040	420	510	31	80
4.3 kJ/mm, AW	0.009	420	510	29	79
	0.018	430	500	29	79
	0.029	420	490	33	79
	0.041	410	500	26	79
	0.049	420	500	28	78
1 kJ/mm, SR	0.007	460	540	32	79
	0.015	460	540	29	79
	0.023	480	550	31	78
	0.030	480	550	29	77
	0.040	480	560	31	76

Note: AW – as-welded; SR – stress relieved.

about 75 and 80% acicular ferrite, 10–15% grain boundary ferrite, the remainder being ferrite with aligned second phase. Increasing heat input increased the amount of primary ferrite, as in the results described in Chapter 4. Experiments to examine the heterogeneity of welds containing phosphorus are described in Chapter 17.

Mechanical properties

Hardness values of the deposits at 1 kJ/mm rose slightly with increasing phosphorus by about 10 HV5 over the whole range.

Tensile strength, yield strength and ductility values appeared to be unchanged by phosphorus in the as-welded condition, although strength rose slightly with phosphorus after stress-relief (Table 6.6). However, regression analysis of the 1 kJ/mm deposits, taking into account the influence of carbon and silicon, showed phosphorus to have a slight strengthening effect in both conditions, as shown in the regression equations developed:

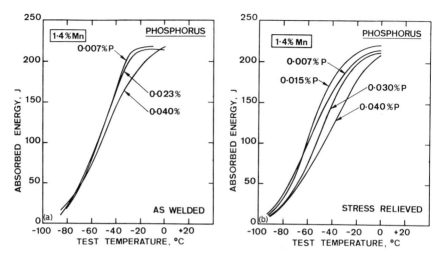

6.19 Influence of phosphorus on Charpy V transition curve: *a)* 1 kJ/mm, as-welded; *b)* 1 kJ/mm, stress-relieved at 580 °C.

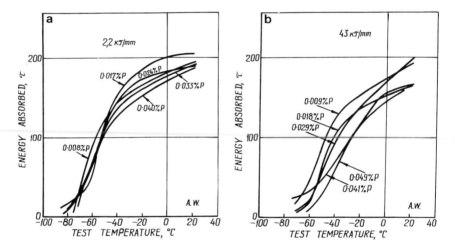

6.20 Influence of phosphorus on Charpy V transition curve: *a)* 2.2 kJ/mm, as-welded; *b)* 4.3 kJ/mm, as-welded.

$$YS_{aw} = 504 + 278 \ P \qquad\qquad [6.11]$$

$$TS_{aw} = 555 + 480 \ P \qquad\qquad [6.12]$$

$$YS_{sr} = 457 + 592 \ P \qquad\qquad [6.13]$$

$$TS_{sr} = 532 + 577 \ P \qquad\qquad [6.14]$$

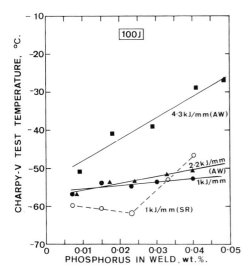

6.21 Influence of phosphorus on 100 J Charpy V transition temperature.

It is notable that the strengthening effect of phosphorus on yield strength in the stress-relieved condition is over twice as great as as-welded.

Charpy curves (Fig. 6.19) from the 1 kJ/mm deposits in the as-welded condition were almost identical except for the 0.040%P deposit, which was slightly less tough. However, after stress relief, phosphorus above about 0.015% led to a distinct deterioration, so that stress relief was slightly beneficial at low phosphorus content and somewhat harmful at the highest level. Increasing heat input gradually increased the spread of values (Fig. 6.20), particularly at high Charpy energies. Figure 6.21 shows that the 100 J Charpy transition temperature increased markedly, particularly over 0.025%P, for both the 4.3 kJ/mm and the stress-relieved 1 kJ/mm deposits.

Arsenic

These tests, previously unpublished, were restricted to as-welded tests on 1.4%Mn deposits with arsenic additions giving up to 0.1%As in the weld metal. The compositions in Table 6.7 show satisfactory levels of all elements.

No metallographic examination was carried out. The mechanical properties, summarised in Table 6.8, show no perceptible effect of arsenic on strength, ductility or toughness.

Table 6.7 Composition of 1.4%Mn weld metal having varying arsenic content

Element, wt%	C	Mn	Si	S	P	As	N	O
As, ppm								
20	0.073	1.52	0.33	0.009	0.014	0.0020	0.080	0.046
290	0.074	1.53	0.31	0.011	0.011	0.029	0.070	0.044
570	0.067	1.48	0.28	0.010	0.012	0.057	0.070	0.048
870	0.065	1.43	0.26	0.010	0.011	0.087	0.070	0.046
1000	0.069	1.45	0.29	0.010	0.012	0.10	0.070	0.046

Note: typical levels of other elements are given in Table 3.1.

Table 6.8 Tensile properties, as-welded, of 1.4%Mn welds of varying arsenic content

As, ppm	Yield strength, N/mm^2	Tensile strength, N/mm^2	Elongation, %	Reduction % of area, %	Charpy transition, °C	
					100 J	28 J
20	470	550	28	78	−71	−91
290	470	550	28	75	−68	−90
570	460	540	30	77	−65	−85
870	460	540	27	78	−62	−87
1000	480	550	28	75	−64	−92

References

1 Evans GM: 'Effect of silicon on the microstructure and properties of C–Mn all-weld-metal deposits'. *Metal Construction* 1986 **18**(7) 438R–44R; *Welding Res Abroad* 1991 **37**(2/3) 20–31; IIW Doc II–A–630–84; *Schweissmitteilungen* 1986 **44**(110) 19–33.

2 Abson DJ: 'A study of the influence of C and Si on the microstructure and toughness of submerged arc welds'. TWI Report, 68/1978/M, July 1978.

3 Evans GM: "Effect of sulphur and phosphorus on the microstructure and properties of C–Mn all-weld-metal deposits'. *Metal Construction* 1986 **18**(9) 631R–36R; *Welding Res Abroad*, 1991 **37**(2/3) 32–41; IIW Doc II–A–640–85; *Schweissmitteilungen* 1986 **44**(111) 22–35.

4 Widgery DJ: 'New ideas in submerged-arc welding'. Proc conf Trends in steels and consumables for welding, London, 1978, TWI.

5 Hart PHM: 'The influence of steel cleanliness on HAZ hydrogen cracking: the present position'. *Welding in the World* 1985 **23**(9/10) 230–40.

6 Steel AC: 'The effects of sulphur and phosphorus on the toughness of mild steel weld metal'. *Welding Res Intl* 1972 **2**(3) 37–76.

7 Pokhodnya IK, Voitkevitch VG, Alexeev AA, Denisenko AV and Evans GM: 'Effect of P on weld impact toughness and chemical heterogeneity'. IIW Doc II–A–823–90.

7

Nitrogen and strain ageing

The major concern regarding nitrogen is its role in promoting strain ageing embrittlement,[1] although nitrogen also acts as an alloying element. The sources of nitrogen are twofold; the welding consumables (wire and raw materials) and contamination from the 80% of nitrogen in the air. This should be kept at bay by CO and other gases generated by decomposition of carbonates and by volatilisation of materials in the electrode coating. In some types of welding, for example submerged-arc, nitrogen contamination from the air is nearly always small (~10 ppm) and virtually predictable. In MMA welding,[2] it depends on the skill and concentration of the welder in maintaining a constant arc length, despite inevitable distractions which can lead to considerable variations in nitrogen content over distances of a centimetre or so. Particular problems are likely in root runs, where atmospheric contamination can come from below and behind, and in striking a new electrode if the correct procedure is not carried out. Excessive nitrogen contents, greater than 250–300 ppm when welding at ambient pressure, can lead to porosity in mild steel weld metal.

Nitrogen as an alloying element

In the tests reported, nitrogen was added up to 200 ppm, using nitrided manganese in the electrode coating to give welds containing 1.25%Mn.[3] The behaviour of nitrogen in high purity weld metals is studied in Chapter 12. The maximum nitrogen level of 200 ppm was short of that which would give porosity. Analyses of the deposits in Table 7.1 show all unvarying elements to be constant. Free nitrogen content, which was estimated from the difference between total and fixed nitrogen, increased as the total nitrogen was increased. Fixed nitrogen content was relatively constant, varying randomly between 9 and 13 ppm.

Table 7.1 Composition of 1.25%Mn weld metal having varying nitrogen content

Element, wt%	C	Mn	Si	Ti	N, total	N, free	O
Nominal N, ppm							
80	0.063	1.24	0.34	0.0045	0.0083	0.0070	0.047
110	0.059	1.22	0.34	0.0044	0.011	0.0104	0.048
150	0.060	1.25	0.37	0.0044	0.015	0.0140	0.046
200	0.060	1.19	0.34	0.0048	0.021	0.0202	0.047

Note: welds contained 0.007%S, 0.008–0.010%P, 0.001%Nb, 0.013–0.014%V, <0.0005%Al, 0.0005–0.0008%B, ~0.027%Ni, ~0.03%Cr, ~0.004%Mo, ~0.024%Cu.

Table 7.2 Effect of nitrogen on microstructure of the as-deposited region of 1.25%Mn welds

Nitrogen, ppm	Ferrite with second phase, %	Acicular ferrite, %	Primary ferrite, %
83	7	78	14
110	11	72	17
150	14	68	17
210	16	65	18

Metallography

Nitrogen had no effect on the macrostructure of the welds examined. In the as-deposited region, nitrogen tended to decrease the size and amount of acicular ferrite and increase that of primary ferrite and ferrite with aligned second phase (Table 7.2). The amount of the last-named constituent increased by almost 10% in the as-welded condition. Examination at high magnification showed that the constituent identified as ferrite with aligned second phase was present in slightly greater proportion (up to 5%) after stress relief. Microhardness measurements are described later, in the section on strain ageing.

Mechanical properties

Tensile properties in the as-welded condition (Table 7.3) showed that strength increased with increasing nitrogen content. Stress relief gave a small, fairly uniform reduction of about 10 N/mm^2 in tensile strength, regardless of nitrogen content but yield strength values were more affected; the drop was 30 N/mm^2 with 80 ppm N and

Table 7.3 Tensile test results on welds of varying nitrogen content

Condition	N, ppm	Yield strength, N/mm^2	Tensile strength, N/mm^2	Elongation, %	Reduction of area, %
AW	83	490	550	30	79
	110	500	560	28	78
	150	530	580	28	76
	210	550	600	27	75
SR	83	460	540	28	78
	110	470	540	32	77
	150	490	560	29	77
	210	500	590	29	76

Note: AW – as-welded; SR – stress relieved.

50 N/mm^2 with 200 ppm. Best fit straight lines (nitrogen contents in wt %) were:

$$YS_{aw} = 445 + 4470\,N \qquad [7.1]$$

$$TS_{aw} = 511 + 4340\,N \qquad [7.2]$$

$$YS_{sr} = 443 + 2660\,N \qquad [7.3]$$

$$TS_{sr} = 501 + 4100\,N \qquad [7.4]$$

Compared with carbon (Chapter 3), nitrogen appears to have a greater effect on strength on a percentage basis for these particular welds.

Charpy tests were carried out in both stress-relieved and as-welded conditions. The results, shown later in the next section, indicated that nitrogen impaired toughness. As-welded, the increase in the 100 J temperature was just over 20 °C on increasing nitrogen from 80 to 210 ppm. Stress relief lowered transition temperatures by an increasing amount as the nitrogen content was increased, so that in the stress-relieved condition, nitrogen was less detrimental than as-welded.

Strain ageing

In high quality steels, it is possible to avoid the problem of strain ageing embrittlement. This is achieved by ensuring that the steel contains sufficient aluminium to combine with all the nitrogen present, and by using a heat treatment schedule that allows the two elements to combine to form aluminium nitride, AlN, and also allows the small amount of free carbon to be combined as carbide. Such precautions are not possible in most weld metals because all contain sufficient nitrogen (i.e. at least 1–2 ppm or 0.0001–0.0002%) to give rise to strain ageing and few contain sufficient free aluminium (i.e. not combined as Al_2O_3) to combine with this nitrogen. Also, cooling is normally too fast, and time at elevated temperature is too short, to allow such combination in the as-deposited condition.

Strain ageing[4] is an interaction between free nitrogen (and free carbon) atoms in the steel and dislocations in the atomic lattice. It can occur at ambient temperature and above – a temperature of about 250 °C is perhaps the most significant at the slow strain rates employed in normal tensile tests. Strain ageing increases strength and reduces ductility and toughness. Heat treatment at a sufficiently high temperature (i.e. stress relieving) removes its effects by giving time for the free nitrogen to combine with elements such as iron and manganese (even when free aluminium is absent), which form nitrides slowly.

When a weld metal cools, it is constrained to deform plastically to accommodate contraction and, in cooling through the strain ageing temperature range (from about 300 down to 150 °C or below), it undergoes dynamic strain ageing. This ageing is most noticeable in the weld root, which is not only likely to pick up more nitrogen from the atmosphere than the rest of the weld but is also likely to be more heavily strained. Ageing of a steel (or weld metal) which has been previously strained is known as static strain ageing.

When a completed weld is strained within the strain ageing range, as in service at moderately elevated temperature, it undergoes dynamic strain ageing. Weld metal tested within the strain ageing temperature range shows higher tensile strength and lower ductility than at ambient temperature.

In devising weld metals to resist strain ageing, several problems would need to be overcome:

(i)   raw materials for welding consumables, particularly welding rod and wire, contain damaging amounts of nitrogen;

(ii)   welding is normally carried out in air and it is virtually impossible to avoid some contamination of nitrogen into the weld metal;

(iii)   weld metals relying on acicular ferrite microstructures to develop toughness require a small content of titanium (probably as an oxide) to be associated with non-metallic inclusions to allow nucleation of the acicular ferrite; addition of sufficient aluminium for it to be in solution (i.e. in a form capable of combining with nitrogen) would limit formation of oxides of titanium, as prior formation of Al_2O_3 would remove oxygen which could otherwise form titanium oxides.

STEP 1 - COLD DEFORMATION (10%)

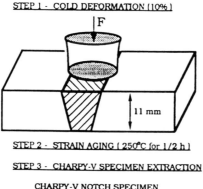

STEP 2 - STRAIN AGING (250°C for 1/2 h)

STEP 3 - CHARPY-V SPECIMEN EXTRACTION

CHARPY-V NOTCH SPECIMEN

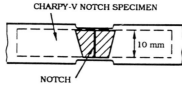

NOTCH

7.1 Procedure employed to produce statically strain aged specimens for Charpy testing.

Some of these factors are taken into consideration when examining selected weld metal additions and ways of keeping nitrogen levels low in the high purity weld metals discussed in Chapter 12.

Welds examined

The welds used to study effects of nitrogen as an alloying element were also used to study static strain ageing in normal C–Mn weld metals.[3] As shown schematically in Fig. 7.1, oversize, as-welded Charpy blanks were compressed 10% at ambient temperature and then aged for 0.5 hour at 250 °C. Specimens were then tested as aged and also after strain ageing followed by a stress relief heat treatment, usually of 2 hours duration at 580 °C. The results were compared with those from as-welded or stress relieved specimens. Similar blanks were used to provide miniature tensile test specimens. The composition of the welds examined had total nitrogen contents ranging from 80 to 210 ppm (Table 7.1).

Microscopic examination at a relatively high magnification (×1000) showed that strain ageing appeared to increase the proportion of ferrite with aligned second phase by between 6.5 and 13% over the amount which was present as-welded (Table 7.2). This was attributed to precipitation of carbides and nitrides around dislocations during

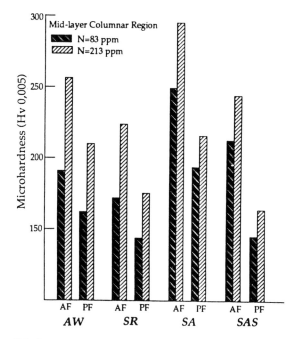

7.2 Summary of microhardness measurements on constituents of 1.25%Mn welds of low and high nitrogen content (AW – as-welded, SA – strain-aged, SAS – strain-aged followed by stress relief, SR – stress-relieved at 580 °C).

strain ageing. Stress relieving after strain ageing added a few more percent of ferrite with aligned second phase, as it did to the as-welded condition.

Microhardness surveys were made on welds of low and high nitrogen content (coded 1 and 4) to assess the influence of strain ageing (and also stress relief) on the acicular ferrite and primary ferrite constituents; it was not possible to examine ferrite with aligned second phase because of the two-phase nature of this constituent. The results, summarised in Fig. 7.2, show that each constituent was hardened by strain ageing and softened by stress-relief, whether or not the latter was applied to an as-welded or a strain-aged deposit. The increase in nitrogen increased microhardness for both primary and acicular ferrite by about 32%.

Strain ageing produced a sharp increase in hardness of both constituents; the increase (both absolutely and as a percentage) was greater for acicular ferrite and for the weld of lower nitrogen content. Stress relieving after strain ageing reduced hardness levels below those

Table 7.4 Tensile test results on 1.2%Mn welds of varying nitrogen content

Condition	N, ppm, total	Yield strength, N/mm²	Tensile strength, N/mm²	Elongation, %	Reduction of area, %
AW	83	490	550	30	79
	113	500	560	28	78
	153	530	580	28	76
	213	550	600	27	75
SA	83	730	750	17	72
	113	720	750	17	73
	153	740	770	17	71
	213	770	790	18	70
SR	83	460	540	28	77
	113	470	540	32	77
	153	490	560	29	77
	213	500	590	29	76
SA/SR	83	640	680	21	71
	113	640	680	20	70
	153	640	690	20	71
	213	660	710	20	68

Note: AW – as-welded; SA – strained 10% compression + aged 0.5 hours at 250 °C; SR – stress relieved 2 hours at 580 °C.

in the as-welded condition, except for the acicular ferrite at the lower nitrogen level. At the higher nitrogen level, the hardness values after strain ageing and stress relieving were similar to those from simply stress-relieved samples; at the lower nitrogen content the stress-relieved values were lower.

Mechanical properties

All welds were tensile tested in the strain aged and the strain aged and stress-relieved conditions as well as as-welded and stress-relieved (given in Table 7.3). The result of all tests are given in Table 7.4 and shown graphically in Fig. 7.3. Linear equations showing the influence of nitrogen on strength properties in the two strain aged conditions are given in Eq. [7.5]–[7.8]; these should be compared with the earlier Eq. [7.1]–[7.4] for the as-welded and the stress-relieved conditions. Strain ageing reduced the linear effect of nitrogen but gave a large increase of nearly 250 N/mm² for the intercept for yield strength. Stress relieving after strain ageing reduced the intercept by over 80 N/mm², and also reduced the linear effect of nitrogen. However, the intercept was 175 N/mm² greater for the strain aged and stress

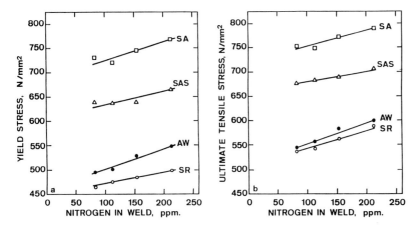

7.3 Effect of nitrogen on strength properties of 1.25%Mn weld metals in different conditions: *a)* Yield strength; *b)* Tensile strength.

relieved condition than for stress relieving without prior strain ageing. Tensile strength values followed a similar pattern:

$$YS_{sa} = 692 + 3520\,N \qquad\qquad\qquad [7.5]$$

$$TS_{sa} = 719 + 3420\,N \qquad\qquad\qquad [7.6]$$

$$YS_{sas} = 618 + 1840\,N \qquad\qquad\qquad [7.7]$$

$$TS_{sas} = 658 + 2230\,N \qquad\qquad\qquad [7.8]$$

Strain ageing also gave a sharp reduction in elongation and a modest fall in reduction-of-area values. Although stress relieving after strain ageing reduced strength and increased ductility, none of the strength or ductility values were restored to the as-welded values, let alone the values after stress relief without prior strain ageing. Additional tests made on strain-aged specimens with 80 ppm N stress-relieved at different temperatures (Fig. 7.4), indicated that a stress relief temperature of about 700 °C would be needed to recover the original stress-relieved tensile strength values (550 N/mm²), although 650 °C would be adequate to revert to the as-welded yield strength level of ~500 N/mm².

Charpy tests, summarised as transition temperatures in Fig. 7.5, showed that, although nitrogen itself slightly increased transition temperature, strain ageing had a much more marked effect, increasing the 100 J temperature by about 50 °C. In addition, strain ageing appeared to reduce upper shelf energy and, at higher nitrogen levels, reduce the transition slope. The results, including tests on welds of different man-

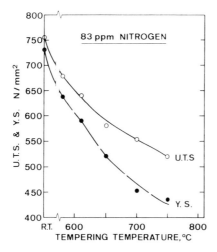

7.4 Effect of tempering temperature on strength properties of strain-aged 1.25%Mn–83 ppm N weld metal.

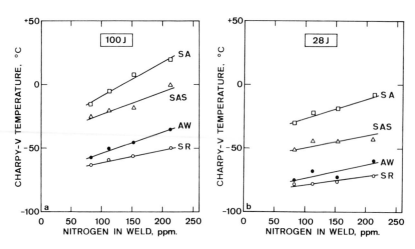

7.5 Effects of nitrogen and strain ageing on Charpy transition temperature for 1.25%Mn weld metal: *a)* 100 J; *b)* 28 J.

ganese content from Chapter 3, are summarised in Table 7.5. Varying manganese or nitrogen had little effect on the increase in the 100 J temperature produced by strain ageing.

Stress relief after strain ageing restored some of the increase in transition temperature, more so for the 28 J than the 100 J temperature,

Table 7.5 Influence of strain ageing on transition temperature of standard C–Mn weld metal

Electrode and compositional type, Mn, %, N ppm	100 J Charpy transition temperature, °C		Increase, °C
	As-welded	Strain aged	
1 1.25 Mn, 80 N	−58	−15	43
2 110 N	−50	−7	43
3 150 N	−45	+7	52
4 210 N	−36	+19	55
A★, 0.6 Mn, 70 N	−27	+5	32
B★, 1.0, 100 N	−44	−5	39
C★, 1.4, 90 N	−53	−12	41
D★, 1.8, 90 N	−43	−19	24

Notes: 10% pre-compression at ambient temperature +0.5 hr/250 °C.
★ For composition see Table 3.1.

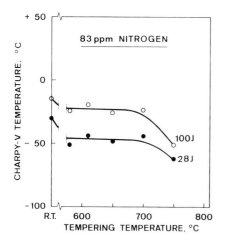

7.6 Effect of tempering temperature on Charpy transition temperature of strain-aged 1.25%Mn–83 ppm N weld metal.

although toughness after strain ageing and stress relief was never as good as in the as-welded or the stress-relieved conditions. Additional tests on the 1.25%Mn deposit with 80 ppm N (Fig. 7.6) to explore higher stress relief temperatures, showed no further improvement in toughness unless the stress relief temperature was increased to 750 °C – an unacceptably high level for practical use.

Table 7.6 Composition of 1.3%Mn weld metal having varying nitrogen content

Element, wt%

C	Mn	Si	N, total	N, interstitial	Interstitial elements*
0.065	1.37	0.39	0.0083	0.0036	0.0045
0.062	1.43	0.41	0.0113	0.0034	0.0043
0.061	1.34	0.41	0.0153	0.0053	0.0066
0.061	1.28	0.37	0.0213	0.0055	0.0069

Notes: welds contained 0.005–0.006%S, 0.006%P, 0.043%O, other elements as Table 3.1.
* Including nitrogen; the interstitial values relate to the as-welded condition.

Damping capacity measurements

Tests were carried out at the Technical University of Delft[5] to assess the amount of free nitrogen (and other atoms) in 1.3%Mn weld metal of four different nitrogen levels (Table 7.6), as-welded and after stress relief at 580 °C.

The results showed that the internal friction measured was due to Snoek damping, free from interstitial damping and cold-working damping. As the total nitrogen content was increased from 83 to 213 ppm, the interstitial nitrogen in the as-welded condition increased from 35 to 55 ppm and the proportion of other interstitial elements (presumably mainly carbon) remained close to 10 ppm (Table 7.6). Stress relief completely removed all interstitial elements. Plastic deformation, in combination with heat treatment at moderate temperature, led to a decrease in concentration of free interstitial nitrogen and to formation of precipitated nuclei.

References

1 Kotecki DJ (ed): 'Nitrogen in arc welding – a review'. *WRC Bull* No. 369, Dec 1991, Jan 1992, Welding Research Council, New York.
2 Morigaki O et al: 'Development of a covered electrode for steel structures in low temperature service'. IIW Doc II–746–75.
3 Kocak M, Achar DRG and Evans GM: 'Influence of SMA weld metal N content on its fracture toughness behaviour'. Proc 2nd int OPEC conf, San Francisco, 1992, pp. 239–47.
4 Baird JD: 'The effects of strain ageing due to interstitial solutes on the mechanical properties of metals'. Metallurgical Review No. 149, *Mat Metals* 1971 5(2) 1–18.
5 den Ouden G and Peekstok ER: 'Internal friction measurement of C–Mn steel weld metal'. IIW Doc II–A–798–90.

Part II

Low alloy steel weld metals

8

Single additions:
with varying manganese

The major alloying elements added to weld metals are chromium, molybdenum and nickel. Also, copper occurs in weathering and HSLA steels, and as an impurity in electrodes and from the copper coating of wires used for automatic welding. Although chromium increases the hardenability of a steel to a similar extent to manganese, its use in weld metals is principally in combination with other elements such as molybdenum for welding steels intended to resist high temperatures, in Ni–Cr–Mo consumables for welding high strength steels and in electrodes for weathering steels.

Molybdenum is not only used with chromium for high temperature applications, but also has an independent role in high strength weld metals with manganese either with or without nickel.

Nickel is the prime element which can increase the toughness of martensite in weld metals[1] and it is often used where toughness is a major property in high and very high strength weld metals, although for some applications requiring resistance to solutions containing H_2S its use in excess of 1% is not recommended.[2]

The work reported in this chapter used weld metals with normal levels of microalloying elements (i.e. capable of giving acicular ferrite in C-Mn weld metals). In all cases test plates were buttered with the consumable under investigation before welding, to avoid dilution problems. Use of high purity weld metals, with or without a small addition of titanium, is considered in Chapters 15 and 16. Otherwise, the experimental techniques were as used in other stages of the project. The alloy-free welds were those whose properties are described under 'Manganese' in Chapter 3.

Chromium (Mn–Cr)

Welds were deposited with chromium contents increasing to 2.3% with the four standard manganese contents of 0.6, 1.0, 1.4 and 1.8%

Table 8.1 Composition of weld metal having varying chromium and manganese content

Element, wt%		C	Mn	Si	Cr
Cr level, %	Electrode code				
0.25	A	0.038	0.65	0.30	0.22
	B	0.041	1.01	0.32	0.24
	C	0.044	1.45	0.32	0.24
	D	0.048	1.85	0.33	0.26
0.5	A	0.040	0.60	0.33	0.51
	B	0.043	0.95	0.31	0.53
	C	0.047	1.42	0.33	0.53
	D	0.051	1.83	0.33	0.52
1.0	A	0.041	0.59	0.30	1.00
	B	0.045	0.97	0.32	1.04
	C	0.048	1.37	0.33	1.08
	D	0.052	1.81	0.33	1.10
2.3	A	0.041	0.59	0.27	2.34
	B	0.046	0.93	0.30	2.38
	C	0.050	1.29	0.31	2.32
	D	0.054	1.72	0.33	2.36

Note: for zero Cr compositions and typical contents of other elements see Table 3.1.

(coded A–D).[3] Their compositions, detailed in Table 8.1, show a slight tendency for carbon to increase but for manganese to decrease, as chromium was increased.

Metallography

Macroscopic examination showed that, as chromium content increased, the distinction between as-deposited and reheated regions became blurred and the general structure appeared to be more columnar (Fig. 8.1). Also apparent in the figure is the tendency (noted in deposits with at least 1%Cr) for the HAZ of the top central bead to be etched darkly, particularly on one side. At low magnification, chromium caused the ferrite veining to disappear and the prior austenite grains could only be faintly discerned by carbide decoration.

Results of point counting of the top, un-reheated beads of deposits with 1 and 1.8%Mn are shown in Fig. 8.2. Up to about 1%, chromium increased the amount of acicular ferrite, mainly at the expense of primary grain boundary ferrite. A maximum amount of

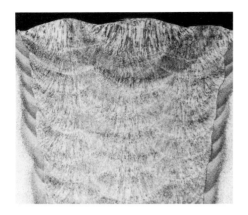

8.1 Macrosection of 2.3%Cr, 1%Mn weld deposit.

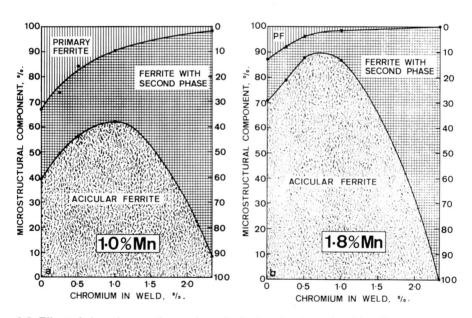

8.2 Effect of chromium on the as-deposited microstructure of welds with:
a) 1%Mn; *b)* 1.8%Mn.

acicular ferrite occurred at about 1%Cr in the deposit with 1%Mn; in 1.8%Mn the maximum acicular ferrite content was between 0.5 and 1%. Further additions of chromium continued to reduce primary ferrite and also acicular ferrite but increased ferrite with aligned second phase, so that 2.3%Cr weld metals contain no primary and little or no acicular ferrite (Fig. 8.3).

Examination at high magnification showed that the acicular ferrite structure was progressively refined up to about 1%Cr, after which it was replaced by ferrite with aligned second phase, which in the 2.3%Cr deposit was partially aligned and partially random, so that the structure was essentially bainitic.

Transmission electron microscopy showed that the microphases, which were of the pearlitic type without chromium (Chapter 3) became of martensite/austenite (M/A) type with 0.25%Cr and became finely dispersed and irregular M/A at 2.3%Cr (Fig. 8.4). Stress relief heat treatment at 580 °C caused extensive carbide precipitation (Fig. 8.5). Selective area diffraction (SAD) showed the precipitates to be predominantly of Fe_3C in the 0.25%Cr deposit but in the 2.3%Cr weld metal the larger precipitates were of $(Fe,Cr)_3C$ and the smaller of chromium nitride, CrN. Precipitation in stress-relieved weld metals with chromium is further discussed in Chapter 17.

With addition of chromium, the high temperature reheated areas lost their ferrite veining and appeared similar to the as-deposited regions, with a fine bainitic structure in which the prior austenite grains could be seen by intermittent carbide decoration. The fine-grained low temperature reheated regions (Fig. 8.6) progressively changed from the equiaxed morphology of the zero and 0.25%Cr deposits to a bainitic structure, whose ferrite grain size could not be determined. Intercritically reheated regions (as in the weld metals containing molybdenum and discussed in that section) showed heavy decoration with carbide.

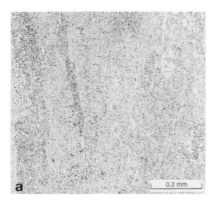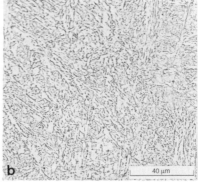

8.3 Microstructure of as-deposited regions of 1%Mn–2.3%Cr weld.

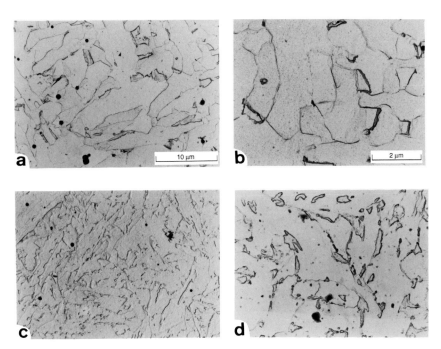

8.4 Microphases in as-deposited regions of deposits with 1%Mn and:
a) 0.25%Cr; *b*) 0.25%Cr; *c*) 2.3%Cr; *d*) 2.3%Cr.

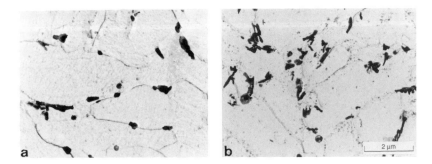

8.5 Carbide precipitation on PWHT in as-deposited regions of welds with
1%Mn and: *a*) 0.25%Cr; *b*) 2.3%Cr.

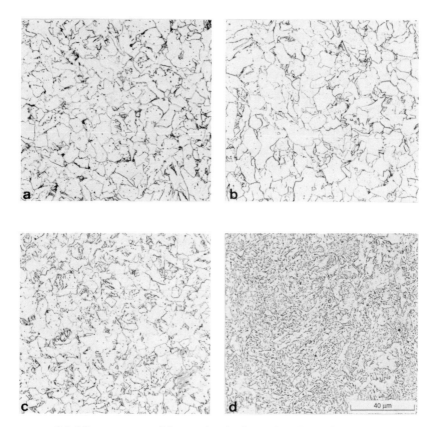

8.6 Microstructure of fine-grained reheated regions of deposits with 1%Mn and: *a*) 0.25%Cr; *b*) 0.5%Cr; *c*) 1%Cr; *d*) 2.3%Cr.

Mechanical properties

Hardness tests on the final, as-deposited runs showed that chromium strongly increased hardness (Fig. 8.7). At the two lower manganese levels the increase was approximately linear with increasing chromium but at the highest manganese level the influence of chromium lessened as its content increased.

Hardness traverses perpendicular to the weld surface (Fig. 8.8), showed a slightly smaller drop in hardness below the top bead as chromium was increased, although this was largely masked by the increasing variability of hardness. The higher chromium weld metals experienced a gradual drop in hardness towards the root of the weld, as a result of tempering.

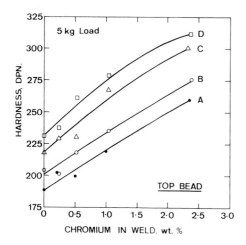

8.7 Effect of chromium on as-deposited hardness of welds with varying manganese content.

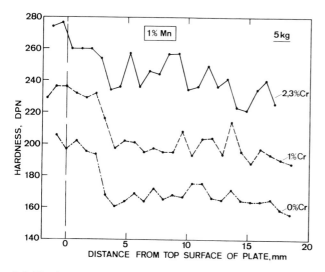

8.8 Hardness traverses of deposits containing 1%Mn and varying chromium content.

Tensile properties from Table 8.2 are plotted in Fig. 8.9. Both yield strength and tensile strength appeared to increase linearly with chromium; regression analysis (Eq. [8.1]–[8.4]) showed weak interactions between Mn and Cr; these were particularly weak after PWHT:

Table 8.2 Tensile test results from weld metal having varying chromium and manganese content

State	Cr, %	Electrode code	Yield strength, N/mm^2	Tensile strength, N/mm^2	Elongation, %	Reduction of area, %
AW	0.25	A	420	500	30	79
		B	460	530	30	78
		C	500	580	28	79
		D	570	620	26	75
AW	0.5	A	450	510	32	79
		B	490	540	30	77
		C	540	600	28	77
		D	620	670	23	70
AW	1.0	A	470	530	26	78
		B	530	590	25	76
		C	630	680	24	71
		D	690	720	22	71
AW	2.3	A	590	650	21	72
		B	660	720	20	68
		C	740	790	20	67
		D	810	860	18	65
SR	0.25	A	410	500	31	80
		B	430	520	30	79
		C	470	560	28	77
		D	540	620	27	74
SR	0.5	A	420	500	29	79
		B	460	540	29	77
		C	510	600	26	76
		D	570	650	24	69
SR	1.0	A	430	510	29	78
		B	470	570	26	77
		C	550	630	26	72
		D	590	670	23	72
SR	2.3	A	530	620	22	75
		B	600	680	20	72
		C	660	730	19	69
		D	700	760	20	70

Notes: for zero Cr tensile results see Table 3.5.
AW – as-welded; SR – stress relieved 2 hr at 580°C.

$$YS_{aw} = 320 + 113 \, Mn + 64 \, Cr + 42 \left(Mn.Cr \right) \qquad [8.1]$$

$$TS_{aw} = 395 + 107 \, Mn + 63 \, Cr + 36 \left(Mn.Cr \right) \qquad [8.2]$$

$$YS_{sr} = 312 + 100 \, Mn + 58 \, Cr + 22 \left(Mn.Cr \right) \qquad [8.3]$$

$$TS_{sr} = 393 + 106 \, Mn + 66 \, Cr + 10 \left(Mn.Cr \right) \qquad [8.4]$$

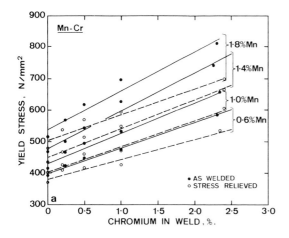

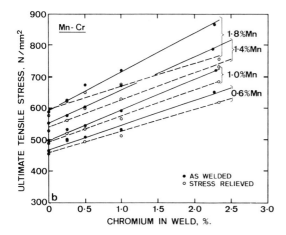

8.9 Effect of chromium and manganese on strength of welds, as-deposited and after stress relief at 580 °C: *a)* Yield strength; *b)* Tensile strength.

Charpy tests (Fig. 8.10(*a*)), showed a relatively small reduction in as-welded toughness with chromium contents up to 1% but a larger fall at 2.3%Cr. The manganese content giving the best toughness fell from about 1.4% with no chromium to just over 1%Mn with 1%Cr and with 2.3%Cr the optimum manganese content was below 0.6% and the 100 J transition temperature above 0 °C. Stress relief at 580 °C (Fig. 8.10(*b*) and 8.11), gave little change in Charpy toughness with deposits containing 0.6 and 1.0%Mn, and with 1.4%Mn and either 0.5 or 2.3%Cr. However, it considerably increased the 100 J

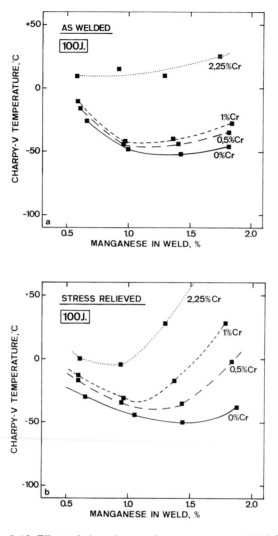

8.10 Effect of chromium and manganese on 100 J Charpy transition temperature of deposits: *a*) As-welded; *b*) After stress relief at 580 °C.

transition temperature with 1.8%Mn at any level of chromium above zero.

Although chromium is a strengthening element, additions up to about 1% tended to increase the amount of acicular ferrite so that the effect on toughness is small. Only when the structure became predominantly bainitic did toughness fall markedly, so that welds with

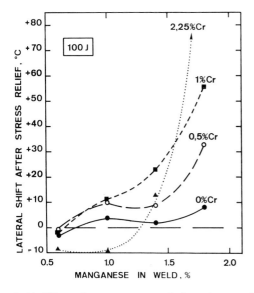

8.11 Effect of manganese and chromium on the shift in 100 J Charpy transition temperature after stress relief at 580 °C.

more than 2%Cr had 100 J transition temperatures above 0 °C, regardless of their manganese content.

Molybdenum (Mn–Mo)

The welds examined[4] had low carbon contents at the four standard manganese levels with molybdenum additions up to 1.1% (Table 8.3). The contents of invariant elements were kept sensibly constant, apart from a tendency for carbon to increase with increasing molybdenum.

Metallography

With 1%Mo, the macrostructure appeared more columnar, as was the case for Cr-containing deposits in the previous section (Fig. 8.1). Point counting of the top, un-reheated runs of deposits containing 1%Mn (Fig. 8.12) showed that up to 0.5%Mo increased the proportion of acicular ferrite at the expense of primary ferrite, although subsequent addition to 1.1%Mo reduced the proportion of both acicular and primary ferrite and increased that of ferrite with aligned second phase. Molybdenum, particularly up to 0.5%, considerably refined the acicular ferrite microstructure (Fig. 8.13). Changes in

Table 8.3 Composition of commercial quality weld metal having varying molybdenum and manganese content

Element, wt%		C	Mn	Si	Mo
Mo level, %	Electrode code				
0.25	A	0.035	0.65	0.33	0.24
	B	0.038	1.03	0.32	0.27
	C	0.044	1.39	0.32	0.25
	D	0.049	1.81	0.34	0.26
0.5	A	0.036	0.63	0.30	0.51
	B	0.039	0.95	0.29	0.52
	C	0.044	1.43	0.33	0.52
	D	0.049	1.78	0.33	0.52
1.1	A	0.042	0.62	0.31	1.11
	B	0.044	1.00	0.34	1.11
	C	0.050	1.37	0.35	1.11
	D	0.051	1.79	0.34	1.12

Note: for zero Mo compositions and typical contents of other elements see Table 3.1.

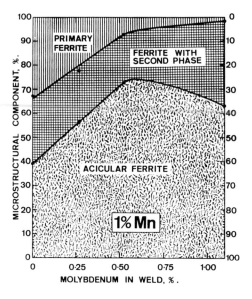

8.12 Effect of molybdenum on as-deposited microstructure of welds with 1%Mn.

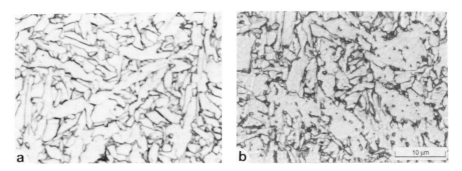

8.13 Microstructure of acicular ferrite in as-deposited regions of 1%Mn welds and: *a*) No Mo; *b*) 0.5%Mo.

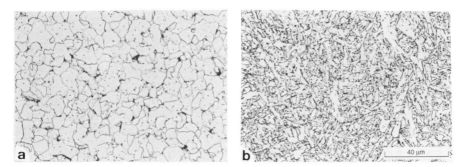

8.14 Microstructure of fine-grained regions of 1%Mn deposits with: *a*) No Mo; *b*) 1.1%Mo.

microstructure of high temperature reheated regions with molybdenum reflected those of as-deposited regions.

The fine-grained regions showed little change with additions up to 0.5%Mo, but with 1.1%Mo the microstructure was not only finer but was predominantly of ferrite with aligned second phase (Fig. 8.14). Electron microscopy showed pearlite and bainite in the deposit with no molybdenum, but with 1.1% the second phase was of the martensite/retained austenite (M/A) type which appeared to change little on stress relief at 580 °C (Fig. 8.15). Intercritically reheated regions of deposits containing molybdenum were marked by a dark-etching grain boundary constituent (Fig. 8.16) which was found to be of Fe_3C and to coincide with the delta-ferrite solidification boundaries.

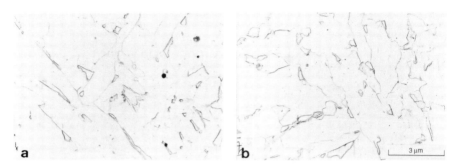

8.15 Electron micrographs of fine-grained regions of 1.1%Mo–1%Mn deposits: *a*) As-welded; *b*) Stress relieved at 580 °C.

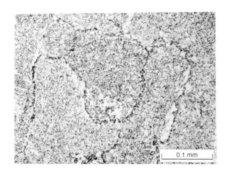

8.16 Intercritically reheated region of 1.1%Mo–1%Mn deposits.

Mechanical properties

Hardness traverses of 1%Mn deposits from the top run downwards (Fig. 8.17) showed the 1.1%Mo weld to be some 40–50 HV5 harder, with more variation in hardness than the Mo-free deposit. The 1.1%Mo deposit gave a less sharp drop in hardness from the top bead to the underlying weld metal than did the corresponding Mo-free weld, whilst the multiply cycled lower weld metal did not soften but tended to harden.

Tensile tests (Table 8.4) revealed a marked strengthening effect, illustrated graphically in Fig. 8.18. Although strengthening was almost linear with respect to both molybdenum and manganese, regression analysis showed a small positive interaction between the two alloying elements which increased on stress-relief:

Table 8.4 Tensile test results from weld metal having varying molybdenum and manganese content

State	Mo, %	Electrode code	Yield strength, N/mm²	Tensile strength, N/mm²	Elongation, %	Reduction of area, %
AW	0.25	A	430	490	28	80
		B	460	530	27	79
		C	540	590	29	78
		D	600	650	26	75
AW	0.5	A	460	540	27	76
		B	500	570	26	76
		C	580	630	25	76
		D	620	680	23	72
AW	1.1	A	550	620	25	74
		B	620	680	24	73
		C	680	730	23	72
		D	720	760	22	71
SR	0.25	A	410	490	32	80
		B	440	520	29	78
		C	500	580	28	75
		D	560	640	28	74
SR	0.5	A	460	550	29	76
		B	520	590	27	75
		C	610	660	25	73
		D	640	700	24	71
SR	1.1	A	580	650	24	72
		B	640	700	23	73
		C	700	770	21	69
		D	730	820	24	72

Notes: for zero Mo tensile results see Table 3.5.
AW – as-welded; SR – stress relieved 2 hr at 580°C.

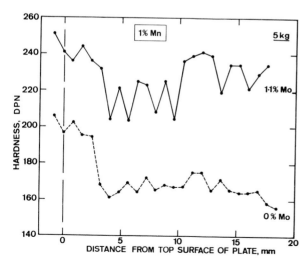

8.17 Hardness traverses of deposits containing 1%Mn, with and without 1.1%Mo.

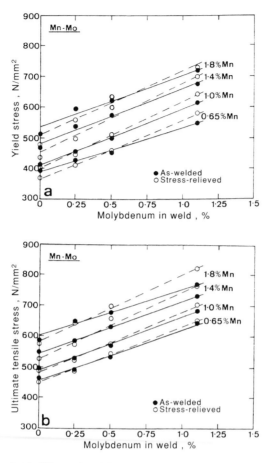

8.18 Effect of molybdenum and manganese on strength of welds as-welded and after stress relief at 580 °C: *a*) Yield strength; *b*) Tensile strength.

$$YS_{aw} = 305 + 121\,Mn + 140\,Mo + 27\left(Mn.Mo\right) \qquad [8.5]$$

$$TS_{aw} = 383 + 116\,Mn + 150\,Mo + 8\left(Mn.Mo\right) \qquad [8.6]$$

$$YS_{sr} = 287 + 113\,Mn + 193\,Mo + 29\left(Mn.Mo\right) \qquad [8.7]$$

$$TS_{sr} = 373 + 113\,Mn + 167\,Mo + 37\left(Mn.Mo\right) \qquad [8.8]$$

From these equations, it appeared that molybdenum had a greater strengthening effect than manganese, particularly after stress relief, so

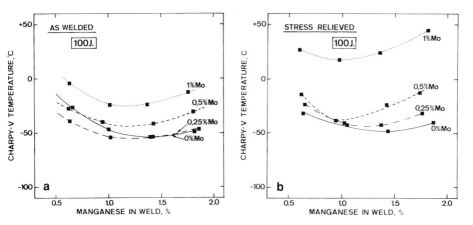

8.19 Effect of molybdenum on 100 J Charpy transition temperature of weld metals of varying manganese level: *a*) As-welded; *b*) Stress-relieved at 580 °C.

that above about 0.5%Mo, stress-relieved welds were stronger than those in the as-welded condition.

As-welded Charpy tests (Fig. 8.19(*a*)) indicated that up to 0.25%Mo was beneficial to the toughness of weld metals containing up to 1%Mn; at 1.4%Mn an addition of 0.25%Mo was harmless; at higher manganese contents, even 0.25%Mo was slightly harmful, Molybdenum contents above 0.25% were generally harmful, so that the 100 J temperatures of 1.1%Mo weld metals were 20–30 °C above those of welds containing no molybdenum.

After stress relief at 580 °C (Fig 8.19(*b*)) molybdenum was generally detrimental to toughness, the increase in transition temperature being particularly marked above 0.5%Mo in weld metals containing up to 1%Mn and above 0.25%Mo for weld metals containing more manganese. The 100 J temperatures were between 15 and 45 °C for stress relieved weld metals with 1.1%Mo.

The results are broadly similar to those of Raiter and Gonzalez,[5] who examined the effect of molybdenum additions up to 0.5% on the as-welded microstructure and properties of weld metals with 1.0, 1.6 and 2.0%Mn.

Nickel (Mn–Ni)

Welds were deposited[6] to give four levels of manganese with additions of 0.5, 1, 2.3 and 3.5%Ni. All elements were held at the desired levels (Table 8.5).

Table 8.5 Composition of weld metal having varying nickel and manganese content

Element, wt %		C	Mn	Si	Ni
Ni level, %	Electrode code				
0.5	A	0.039	0.66	0.32	0.53
	B	0.041	1.01	0.33	0.49
	C	0.051	1.40	0.32	0.47
	D	0.049	1.85	0.33	0.51
1	A	0.038	0.63	0.31	1.09
	B	0.043	1.00	0.33	1.10
	C	0.049	1.37	0.35	1.06
	D	0.053	1.83	0.35	1.06
2.25	A	0.041	0.62	0.30	2.38
	B	0.044	0.96	0.31	2.38
	C	0.049	1.41	0.32	2.32
	D	0.046	1.81	0.32	2.32
3.5	A	0.037	0.65	0.30	3.50
	B	0.041	0.98	0.31	3.46
	C	0.048	1.40	0.33	3.47
	D	0.051	1.79	0.36	3.42

Note: for zero Ni composition and typical contents of other elements see Table 3.1.

Metallography

Nickel, like chromium and molybdenum, blurred the visual signs of the build-up of successive weld runs in macrosections. Unlike these elements, nickel did not enhance the underlying prior austenite grain structure although, like manganese, it did increase the tendency for segregational banding at weld ripple solidification fronts.

Nickel generally increased the proportion of acicular ferrite at the expense of both primary ferrite and ferrite with second phase (Fig. 8.20). At the highest manganese level of 1.8%, nickel above 2.2% gave rise to martensite and an increased proportion of ferrite with second phase at the expense of acicular and primary ferrite. Nickel also increased the aspect ratio of the acicular ferrite (Fig. 8.21 and 8.22), and suppressed cementite films and pearlite with retention of M/A constituent which, after stress relief, precipitated as grain boundary carbides (Fig. 8.23).

Coarse-grained, high temperature reheated regions followed the pattern of the as-welded regions but, unlike the previous alloyed deposits, primary ferrite was not eliminated so that the coarse appearance was maintained (Fig. 8.24). The fine uniform structure of the

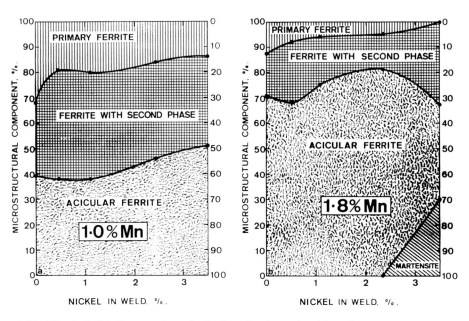

8.20 Effect of nickel on as-deposited microstructure of welds with:
a) 1%Mn; *b*) 1.8%Mn.

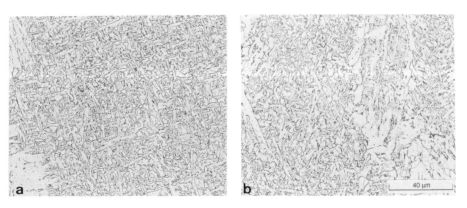

8.21 Microstructure of acicular ferrite in as-deposited regions of 1%Mn welds with: *a*) 0.5%Ni; *b*) 3.5%Ni.

low temperature reheated regions was modified with increasing nickel (Fig. 8.25) as the ferrite was replaced by colonies of ferrite with second phase. This was particularly pronounced at the highest manganese and nickel levels (Fig. 8.26), making it impossible to estimate

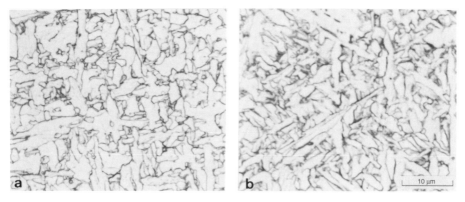

8.22 Microstructure of acicular ferrite in as-deposited regions of 1%Mn welds with: *a*) 0.5%Ni; *b*) 3.5%Ni, TEM replicas.

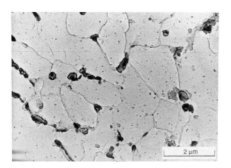

8.23 Microstructure of acicular ferrite in stress relieved weld metal with 1%Mn, 0.5%Ni; TEM replica.

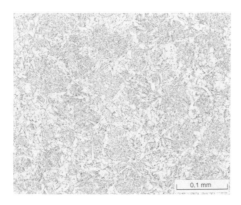

8.24 Microstructure of coarse-grained high temperature region of weld metal with 1%Mn–3.5%Ni.

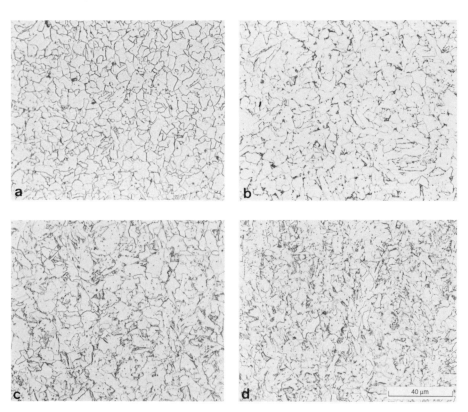

8.25 Microstructure of fine-grained low temperature regions of weld metal with 1%Mn and: *a*) 0.5%Ni; *b*) 1%Ni; *c*) 2.2%Ni; *d*) 3.5%Ni.

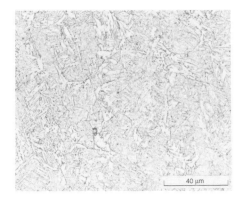

8.26 Microstructure of fine-grained low temperature region of weld metal with 1.8%Mn and 3.5%Ni.

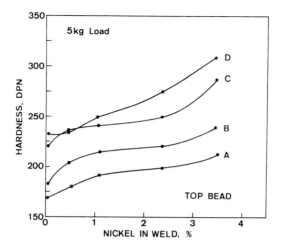

8.27 Effect of nickel on as-deposited hardness of welds with varying manganese content.

the ferrite grain size. Unlike the earlier alloy systems, dark-etching filamentary carbides were not found.

Mechanical properties

Nickel increased hardness in a non-linear fashion in the as-deposited top beads (Fig. 8.27). Although manganese appeared to have an increasing effect between 0.6 and 1.4%, the increase to 1.8% did not produce a large increase in hardness, even at the highest nickel level when martensite was present.

Hardness traverses down selected welds with 1%Mn (Fig. 8.28), showed a drop in hardness below the first bead and an increasing tendency for hardness to fall along the traverse away from the top bead as the nickel content was increased.

Tensile tests (Table 8.6) gave relationships between nickel content and strength which were almost linear (Fig. 8.29), although regression analysis revealed weak interactions between nickel and manganese:

$$YS_{aw} = 332 + 99 \text{ Mn} + 9 \text{ Ni} + 21 \left(\text{Mn.Ni}\right) \quad\quad [8.9]$$

$$TS_{aw} = 401 + 102 \text{ Mn} + 16 \text{ Ni} + 15 \left(\text{Mn.Ni}\right) \quad\quad [8.10]$$

Table 8.6 Tensile test results from weld metal having varying nickel and managnese content

State	Ni, %	Electrode code	Yield strength, N/mm^2	Tensile strength, N/mm^2	Elongation, %	Reduction of area, %
AW	0.5	A	410	480	33	79
		B	460	520	30	78
		C	510	570	29	77
		D	550	620	26	75
AW	1	A	420	500	31	79
		B	480	540	30	78
		C	500	570	28	79
		D	570	640	26	76
AW	2.25	A	470	540	31	78
		B	490	570	31	77
		C	570	640	24	73
		D	620	680	24	74
AW	3.5	A	480	560	30	77
		B	520	600	28	76
		C	620	680	25	73
		D	690	750	22	70
SR	0.5	A	400	480	34	81
		B	430	510	32	80
		C	450	540	28	79
		D	530	610	27	73
SR	1	A	410	490	34	80
		B	440	520	32	78
		C	470	560	30	76
		D	520	610	26	74
SR	2.25	A	430	520	32	77
		B	480	560	29	77
		C	530	620	27	73
		D	570	660	26	71
SR	3.5	A	480	560	31	79
		B	520	600	28	74
		C	580	680	27	71
		D	660	740	24	68

Notes: for zero Ni tensile results see Table 3.5.
AW – as-welded; SR – stress relieved 2 hr at 580°C.

$$YS_{sr} = 319 + 85\ Mn + 17\ Ni + 21\ (Mn.Ni) \qquad [8.11]$$

$$TS_{sr} = 393 + 95\ Mn + 17\ Ni + 19\ (Mn.Ni) \qquad [8.12]$$

Nickel was a much less potent strengthener of weld metal than manganese, chromium or molybdenum. All deposits were weakened by

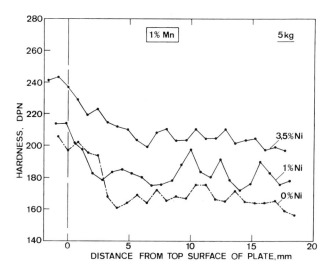

8.28 Hardness traverses down centreline of deposits containing 1%Mn with different nickel content.

stress relief heat treatment, yield strength being the more greatly affected. Slight falls in ductility were also apparent, particularly at high levels of both elements.

Charpy tests are summarised as 100 J transition temperatures in Fig 8.30. Deposits with 0.65%Mn showed a small but steady improvement in toughness as the nickel content was increased in both as-welded and stress relieved conditions, the reduction in the as-welded transition temperature being about 30 °C from zero to 3.4%Ni. With increasing manganese content, nickel became slightly detrimental, as it was above about 1%Ni when 1.4%Mn was present. With 1.8%Mn, toughness continuously deteriorated from 0.5%Ni upwards, so that the 3.5%Ni–1.8%Mn deposit had a 100 J temperature as high as normal ambient temperature. A sharp increase in the rate of deterioration of toughness in the 1.8%Mn deposits occurred above 2.2%Ni, where the as-deposited microstructure (Fig. 8.20(b)) started to contain martensite and an increasing proportion of ferrite with second phase.

In the stress relieved condition (Fig. 8.30(b)), the detrimental effect of nickel was even more marked, so that deposits with 3.5%Ni–1.4%Mn and 2.2%Ni–1.8%Mn both had 100 J transition temperatures close to 60 °C. Nevertheless, in the PWHT condition, additions up to 3.5%Ni were beneficial at 0.65%Mn and were tolerable at 1%Mn.

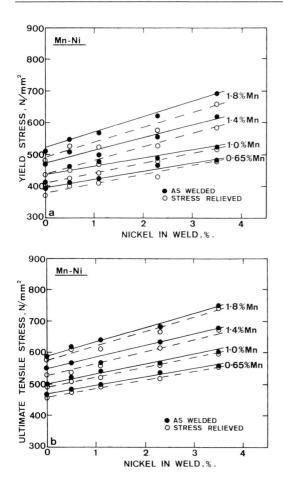

8.29 Effect of nickel and manganese on strength of welds, as-deposited and after stress relief at 580 °C: *a*) Yield strength; *b*) Tensile strength.

Copper (Cu)

Tests were confined to deposits with 1.5%Mn;[7] these contained satisfactory levels of all elements (Table 8.7). However, weld metal carbon contents, averaging 0.074%, were appreciably higher than previous welds with additions of Cr, Mo and Ni.

Metallography

Copper refined grain size in several ways: it refined the columnar grains, prior austenite grain size in the coarse-grained high

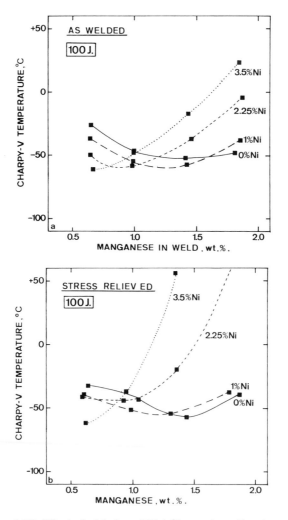

8.30 Effect of nickel on 100 J Charpy transition temperature of weld metal of varying manganese level: a) As-welded; b) Stress relieved at 580 °C.

temperature region, fine-grained reheated weld metal and acicular ferrite in the as-deposited microstructure. These are illustrated for the extreme copper contents in Fig. 8.31. Most refinement of columnar and prior austenite grains occurred with the first 0.2% of copper added (Fig. 8.32). Most refinement in the fine grained weld metal was with the final addition to 1.4%.

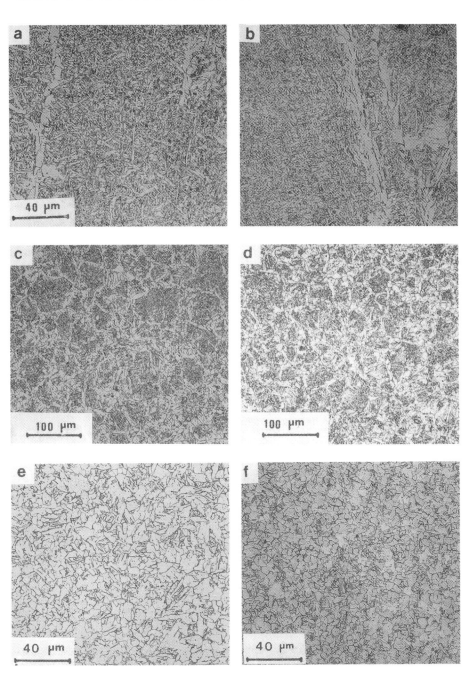

8.31 As-welded structure of 1.5%Mn welds with (a,c,e) 0.02%Cu and (b,d,f) 1.4%Cu showing: a,b) As-deposited; c,d) Coarse-grained reheated; e,f) Fine-grained reheated regions.

Table 8.7 Composition of 1.5%Mn weld metal having varying copper content

Element, wt%	C	Mn	Si	Cu	O
Nominal Cu, %					
0	0.074	1.51	0.36	0.02	0.040
0.1	0.075	1.48	0.36	0.11	0.039
0.2	0.074	1.42	0.32	0.19	0.038
0.35	0.076	1.48	0.33	0.35	0.040
0.7	0.068	1.49	0.32	0.66	0.039
1.4	0.076	1.49	0.36	1.40	0.041

Note: for typical contents of other elements see Table 3.1.

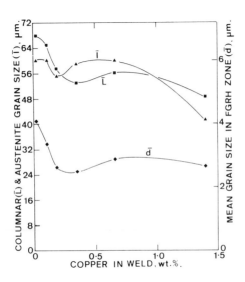

8.32 Effect of copper on grain size parameters of 1.5%Mn weld metal: mean columnar grain width (L̄), mean prior austenite grain size in high temperature reheated weld metal (ī) and mean grain size in fine-grained low temperature reheated weld metal (d̄) (note change of scale in the last).

As well as refining, copper influenced the type of as-deposited microstructure (Fig. 8.33(*a*)). The first 0.1%Cu had a negligible effect but the next 0.1% reduced the amount of acicular ferrite and increased primary ferrite. From 0.2 to 0.65%, copper increased the proportion of ferrite with second phase, mainly at the expense of primary ferrite. Over the whole range, copper increased the proportion of microphases (pearlite, retained austenite, martensite and

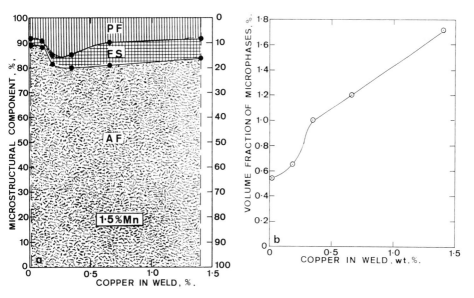

8.33 Effect of copper on 1.5%Mn weld metal microstructure (results of point counting): *a*) As-deposited weld metal, ×630; *b*) Microphases in fine-grained reheated weld metal, SEM.

bainite) in the fine grained reheated zone (Fig. 8.33(*b*)). Some fine carbides were also seen, particularly at the highest copper level.

Stress relief increased the amount of such carbides, which occurred in two distinct types: precipitation along ferrite laths in ferrite with aligned second phase (Fig. 8.34(*a*)), and as isolated groups of fine particles (Fig. 8.34 (*a* and *b*)). The latter are thought to result from tempering of martensite on stress relief. Examination in the TEM gave further detail of copper precipitation and also of non-metallic inclusions, see Chapter 17.

Mechanical properties

Up to 0.2%Cu had no effect on hardness but subsequent additions increased hardness. For copper contents up to 0.7%, stress relief softened the top, as-deposited weld beads and hardened the underlying beads only of the 0.7%Cu weld metal. With 1.4%Cu, both top and underlying beads were hardened by PWHT.

Linear regression equations (Eq. [8.13]–[8.16]) were developed to describe the relatively small strengthening effects of copper in weld

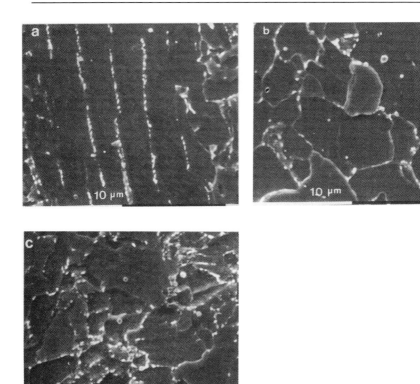

8.34 Carbides in stress relieved 1.5%Mn weld metals: *a*) Between laths in the ferrite with aligned second phase structure with 0.02%Cu; *b*) Isolated groups in the fine-grained region of the same weld metal; *c*) In the fine-grained region of weld metal with 1.4%Cu, SEM micrographs.

metals with 1.5%Mn, in both as-welded and stress relieved conditions (Table 8.8):

$$YS_{aw} = 484 + 57 \ Cu \qquad\qquad [8.13]$$

$$TS_{aw} = 562 + 58 \ Cu \qquad\qquad [8.14]$$

$$YS_{sr} = 472 + 69 \ Cu \qquad\qquad [8.15]$$

$$TS_{sr} = 531 + 107 \ Cu \qquad\qquad [8.16]$$

Figure 8.35 shows that stress relief weakened deposits with low copper content and strengthened those with high copper. The cross-over points were just below 0.1%Cu for tensile strength and just above 0.1%Cu for yield strength.

Table 8.8 Tensile test results from 1.5%Mn weld metal having varying copper content

State	Cu, %	Yield strength, N/mm²	Tensile strength, N/mm²	Elongation, %	Reduction of area, %
AW	0	480	560	26	79
	0.1	500	580	28	77
	0.2	490	570	28	77
	0.35	500	590	28	77
	0.7	510	590	29	77
	1.4	570	650	27	73
SR	0	470	560	31	77
	0.1	460	550	30	78
	0.2	450	550	31	79
	0.35	480	570	28	78
	0.7	500	610	29	74
	1.4	580	670	24	73

Note: AW – as-welded; SR – stress relieved 2 hr at 580°C.

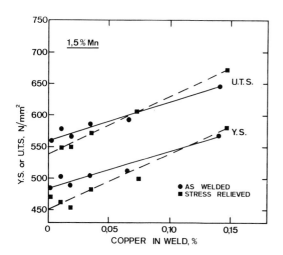

8.35 Effect of copper on strength properties of weld metals with 1.5%Mn.

The results of Charpy toughness tests are presented in terms of 100 J transition temperatures in Fig. 8.36. Additions of up to 1.4%Cu had a mild embrittling effect on as-welded toughness. Up to 0.2% Cu was harmless in either as-welded or stress-relieved conditions. Furthermore, stress relief at 580 °C had no effect on the transition

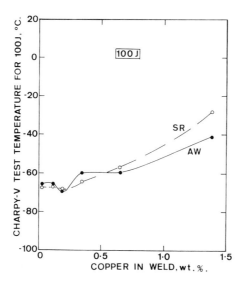

8.36 Effect of copper on 100 J Charpy transition temperature of 1.5%Mn weld metal.

temperature up to 0.7%, although 1.4%Cu was slightly detrimental after stress relief.

The relative influence of the major alloying elements is re-assessed in Chapter 15, in the context of high purity weld metals.

References

1 Saunders GG: 'Effects of major alloying elements on the toughness of high strength weld metal'. TWI report M/86/75, 1975.

2 NACE Standard MR0175-92: 'Standard recommended practice for sulfide stress corrosion-resistant materials for oilfield equipment'. NACE, Houston Tx, 1992.

3 Evans GM: 'Effect of chromium on the microstructure and properties of C–Mn all-weld-metal deposits'. *Welding and Metal Fabrication* 1989 **57**(7) 346s–58s; *Welding Res Abroad* 1991 **37**(2/3) 56–69; IIW Doc II–A–739–88; *Schweissmitteilungen* 1989 **47**(120) 17–34.

4 Evans GM: 'Effect of molybdenum on the microstructure and properties of C–Mn all-weld-metal deposits'. *Joining and Materials* 1988 **1**(15) 239s–46s; *Welding Res Abroad* 1991 **37**(2/3) 42–55; IIW Doc II–A–666–86; *Schweissmitteilungen* 1987 **45**(115) 10–25.

5 Raiter V and Gonzalez JC: 'Influence of molybdenum on the microstructure and properties of C–Mn all-weld-metal with different manganese contents'. *Canad Met Quarterly* 1989 **28**(2) 179–85.

6 Evans GM: 'Effect of nickel on the microstructure and properties of C–Mn all-weld-metal deposits'. *Joining Sciences* 1991 **1**(1) 2s–13s; *Welding Res Abroad* 1991 **37**(2/3) 70–83; IIW Doc II–A–791–89; *Schweissmitteilungen* 1990 **48**(122) 18–35.

7 Evans GM: 'Microstructure and mechanical properties of copper-bearing MMA C–Mn weld metal'. *Weld J* 1991 **70**(3) 80s–90s; *Welding Res Abroad* 1991 **37**(2/3) 84–95; IIW Doc II–A–767–89; *Schweissmitteilungen* 1990 **48**(123) 15–31.

9

Elements in combination – Cr–Mo

In the previous chapter, effects of single alloying additions to C–Mn weld metals were examined. The number of possible permutations of such elements is large and the present chapter concentrates on different combinations of chromium and molybdenum. Over the years, steels for high temperature use in power stations and petrochemical plant have become established as steels of increasing chromium content, namely 0.3 or 0.5%Mo without chromium, 1 or 1.25%Cr with 0.5%Mo, 2.25%Cr with 1%Mo and then, via less frequently used steels of intermediate chromium content, to 9%Cr–1%Mo and 12%Cr–1%Mo, which are outside the scope of this monograph. Consumables for welding Cr–Mo steels have generally been of matching composition, although usually with lower carbon contents.

This chapter, in three parts, examines how the following Cr–Mo weld metals are built up from the simpler weld metals described in the previous chapter. These weld metals (with less than 1%Mn) are probably best known by their American AWS designations for the electrodes:

(i)   1.25%Cr–0.5%Mo (AWS A 5.5 E8018–B2L type) with a low carbon content and the four standard manganese levels used elsewhere in this monograph in four conditions of heat treatment;

(ii) 2.25%Cr–1%Mo weld metals with a low carbon content (AWS A 5.5 E9018–B3L);

(iii) 2.25%Cr–1%Mo weld metals with a higher carbon content (AWS A 5.5 E9018–B3).

Most of the hitherto unreported test programme was similar to what has been described in earlier chapters, except that stress relief heat treatments were more in line with those used when fabricating the type of steel involved and the weld preparations were buttered with the electrodes under test.

1.25%Cr–0.5%Mo weld metals

Weld metals having the compositions given in Table 9.1 were tested and their properties compared with those of C–Mn, Mn–Cr and Mn–Mo deposits of similar composition.

Table 9.1 Composition of 1.25%Cr–0.5%Mo (E8018–B2L type) weld metal having varying manganese content

Element, wt%

C	Mn	Si	S	P	Cr	Mo
0.047	0.67	0.28	0.008	0.009	1.12	0.55
0.050	1.01	0.32	0.007	0.010	1.16	0.53
0.053	1.49	0.34	0.007	0.012	1.16	0.51
0.056	1.92	0.34	0.007	0.014	1.16	0.53

Note: for typical impurity levels, see Table 3.1.

Table 9.2 Tensile test results from 1.25%Cr–0.5%Mo (E8018–B2L type) welds of varying manganese content

Mn, %	State	Yield strength, N/mm^2	Tensile strength, N/mm^2	Elongation, %	Reduction of area, %
0.6	AW	580	650	23	72
1.0		640	710	21	71
1.4		710	750	22	70
1.8		780	830	22	68
0.6	SR	480	570	23	74
1.0		520	590	26	75
1.4		550	620	23	73
1.8		560	640	23	69
0.6	N + T	300	420	39	83
1.0		320	440	34	83
1.4		350	470	34	80
1.8		370	540	30	77
0.6	Q + T	390	490	33	82
1.0		410	520	29	80
1.4		430	550	27	79
1.8		450	570	27	78

Note: AW – as-welded; N + T – normalised and tempered at 600 °C; Q + T – quenched and tempered at 600 °C; SR – stress relieved, 15 hr at 700 °C.

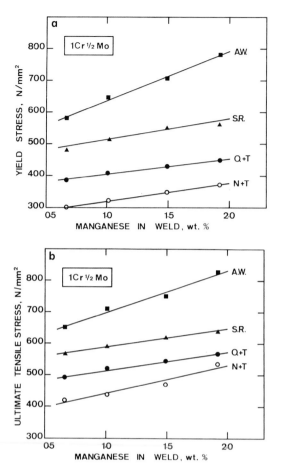

9.1 Effect of manganese on strength of 1%Cr–0.5%Mo weld metal in different heat treatment conditions: a) Yield strength; b) Tensile strength.

Mechanical properties

Tensile properties after different heat treatments are given in Table 9.2. In all conditions, manganese increased strength, usually at the expense of ductility. Post weld heat treatment of 15 hr at 700 °C (typical of what is used on fabrications for high temperature service) was longer in duration and higher in temperature than the 2 hours at 580 °C stress relief used previously for all welds, including those with separate additions of Cr and Mo described in the previous chapter. The higher temperature led to a sharper drop in strength on PWHT.

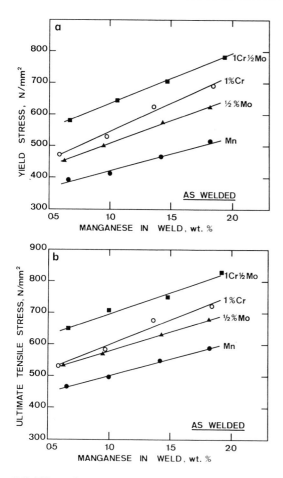

9.2 Effect of manganese on as-welded strength of 1%Cr–0.5%Mo weld metal in comparison with its constituent, simpler weld metals: *a*) Yield strength; *b*) Tensile strength.

Normalising and tempering and quenching and tempering both resulted in much lower strengths than did either the as-welded or the stress-relieved conditions, normalised welds being the softest.

The strengthening effect of manganese was approximately linear for all heat treatments (Fig. 9.1) and the following regression equations describe the behaviour:

$$YS_{AW} = 478 + 157 \text{ Mn} \qquad [9.1]$$
$$YS_{SR} = 448 + 67 \text{ Mn} \qquad [9.2]$$

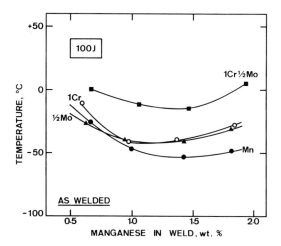

9.3 Effect of manganese on 100 J Charpy transition temperature of 1%Cr–0.5%Mo weld metal in comparison with its constituent, simpler weld metals.

$$YS_{NT} = 260 + 59 \ Mn \qquad\qquad\qquad [9.3]$$

$$YS_{QT} = 356 + 49 \ Mn \qquad\qquad\qquad [9.4]$$

$$TS_{AW} = 563 + 134 \ Mn \qquad\qquad\qquad [9.5]$$

$$TS_{SR} = 530 + 59 \ Mn \qquad\qquad\qquad [9.6]$$

$$TS_{NT} = 350 + 91 \ Mn \qquad\qquad\qquad [9.7]$$

$$TS_{QT} = 454 + 60 \ Mn \qquad\qquad\qquad [9.8]$$

Manganese had a greater strengthening effect as-welded than in any of the other heat treatment conditions (Fig. 9.2). Comparison with Eq. [3.1]–[3.4], [5.5] and [5.6] reveals that the effect of manganese is greater in the presence of 1.25%Cr–0.5%Mo than in its absence in each of the heat treatment conditions tested, despite the difference in PWHT conditions.

Toughness comparisons with the simpler weld metals, using as-welded 100 J transition temperatures, are given in Fig. 9.3. Although the Cr–Mo weld metal was the least tough of those depicted, manganese was beneficial up to 1.4%. After heat treatment (Fig. 9.4), toughness was improved, quenching and tempering being the favoured option, giving 100 J temperatures below −40 °C, almost regardless of manganese content. Stress relieving at 700 °C and normalising and tempering resulted in intermediate toughness, with an optimum manganese content of about 1.2%. The only exception was 1.8%Mn weld metal, where normalising and tempering gave toughness no better than as-welded.

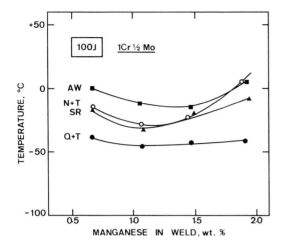

9.4 Effect of manganese on 100 J Charpy transition temperature of 1%Cr–0.5%Mo weld metal in different heat treatment conditions.

Low carbon 2.25%Cr–1%Mo weld metals

The test programme was similar to that for the 1.25%Cr–0.5%Mo weld metals, except that, as 2.25%Cr–1%Mo steels and weld metals can suffer from temper embrittlement,[1] the risk of this problem was assessed by including a step cooling heat treatment after PWHT at 690 °C. No tests were carried out in the quenched and tempered and the normalised and tempered conditions. The step cooling treatment, of approximately 356 hours duration before the final cooling, was as follows (all heating and cooling, except the last, were at 5.5 °C/hr):

heat to 595 °C,
soak for 1 hr at 595 °C,
cool to 540 °C,
soak for 15 hr at 540 °C,
cool to 525 °C,
soak for 24 hr at 525 °C,
cool to 495 °C,
soak for 60 hr at 495 °C,
cool to 470 °C,
soak for 100 hr at 470 °C,
cool to 315 °C,
cool to ambient temperature in still air.

Step cooling was also applied to weld metal with 1%Mn after PWHT at a range of temperatures between 600 and 750 °C.

Table 9.3 Composition of low carbon 2.25%Cr–1%Mo (E9018–B3L type) weld metal

Element, wt%

C	Mn	Si	S	P	Cr	Mo
0.044	0.63	0.26	0.006	0.009	2.40	0.99
0.045	0.89	0.32	0.007	0.008	2.40	1.00
0.045	1.28	0.34	0.006	0.009	2.42	0.94
0.047	1.69	0.33	0.005	0.009	2.40	0.96

Note: for typical impurity levels, see Table 3.1.

Table 9.4 Tensile test results from low carbon 2.25%Cr–0.5%Mo (E9018–B3L type) welds of varying manganese content

Mn, %	State	Yield strength, N/mm^2	Tensile strength, N/mm^2	Elongation, %	Reduction of area, %
0.6	AW	730	800	20	66
1.0		750	830	20	64
1.4		820	890	19	65
1.8		860	900	21	65
0.6	SR	710	840	20	66
1.0	580 °C	720	850	19	66
1.4		740	880	19	66
1.8		780	890	19	66
0.6	SR	470	570	22	75
1.0	690 °C	480	580	25	76
1.4		460	580	23	71
1.8		250	480	29	76
0.6	STC	360	500	27	76
1.0		300	480	30	75
1.4		200	430	35	77
1.8		220	460	32	75

Note: AW – as-welded; SR – stress relieved: either 2 hr at 580 °C or 20 hr at 690 °C, STC – stress relieved 20 hr at 690 °C and then given step cooling heat treatment.

The composition of the weld metals tested is summarised in Table 9.3 and their tensile properties in Table 9.4.

Manganese strengthened only in the as-welded condition and after PWHT at 580 °C (Table 9.4). Stress relief at 690 °C reduced the strengthening effect of manganese almost to zero – in fact the weld

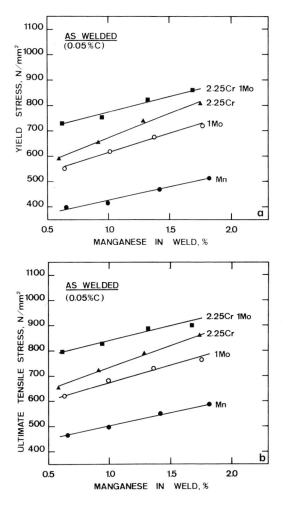

9.5 Effect of manganese on as-welded strength of low carbon 2.25%Cr–1%Mo weld metal in comparison with its constituent, simpler weld metals: *a)* Yield strength; *b)* Tensile strength.

of highest manganese content was significantly weakened, probably because 690 °C is above its A_{c1} temperature and the weld metal was therefore partially transformed to austenite, destroying its dislocation network still present after the 580 °C stress relief. All welds were considerably weakened by step cooling, except for the deposit of highest manganese content, which was already weakened by heating above its A_{c1} temperature.

Table 9.5 Effect of PWHT temperature on tensile properties of low carbon 2.25%Cr–0.5%Mo–1%Mn (E9018–B3L type) weld metal before and after step cooling heat treatment

PWHT temperature, °C	Step Cooling	Yield stress N/mm²	Tensile stress N/mm²	Elongation, %	Reduction of area, %
600	No	680	780	18	69
630		630	730	20	69
660		530	630	23	72
690		480	610	26	71
720		390	540	31	73
750		290	490	33	76
600	Yes	650	760	21	69
630		620	732	20	69
660		570	680	21	70
690		420	550	30	74
720		380	540	28	73
750		310	480	31	72

Note: PWHT time – 20 hr; for step cooling treatment, see text above.

Comparison with data in Chapter 8 on the as-welded strengths of the simpler weld metals from which the Cr–Mo deposits were derived (Fig. 9.5), shows that (as with the 1%Cr–0.5%Mo deposits in Fig. 9.2) the strength of the 2.25%Cr–1%Mo welds was less than would be expected from the separate increments due to Cr and Mo. Unlike the welds of lower chromium content, the strengthening effect of manganese was reduced.

Table 9.5 and Fig. 9.6 show the effects of PWHT temperature on the tensile properties of the low carbon 2.25%Cr–1%Mo weld metal with 1%Mn. Increasing the temperature of PWHT resulted in a steady drop in strength, coupled with an increase in ductility; surprisingly, step cooling had no significant effect on these properties. Although yield strength fell by almost 400 N/mm² as the PWHT temperature was increased from 600 to 750 °C, the reduction in tensile strength was only 300 N/mm² over the same range.

Charpy toughness, expressed as 100 J transition temperature, was poor as-welded (Fig. 9.7). Even with the optimum manganese content of 1–1.5%, the 100 J temperature was almost 50 °C, nearly 30 °C higher than the simpler 2%Cr weld metal. Toughness with 1%Mn was improved by PWHT (Fig. 9.8), the 100 J temperature falling to –30 °C or better after heat treatment between 660 and 720 °C and then increasing slightly with the 750 °C treatment.

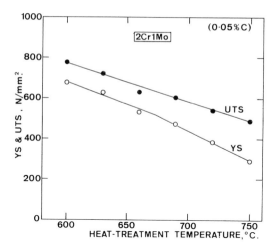

9.6 Effect of PWHT temperature on strength of low carbon 2.25%Cr–1%Mo–1%Mn weld metal.

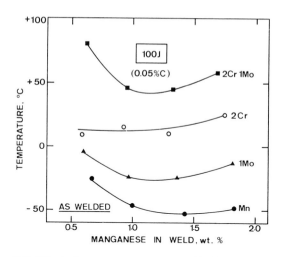

9.7 Effect of manganese on as-welded 100 J Charpy transition temperature of low carbon 2.25%Cr–1%Mo weld metal in comparison with its constituent, simpler weld metals.

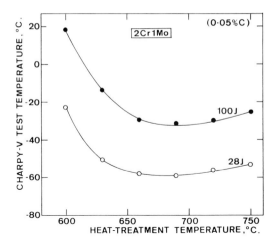

9.8 Effect of PWHT temperature (15 hr treatment) on 100 J Charpy transition temperature of low carbon 2.25%Cr–1%Mo–1%Mn weld metal.

Higher carbon 2.25%Cr–1%Mo weld metals

The composition of these weld metals with almost 0.1%C is given in Table 9.6. Tests were carried out after the same heat treatments used for the lower carbon version of the same weld metal, except that PWHT at different temperatures was not investigated.

Tensile properties in Table 9.7 and Fig. 9.9 show appreciably higher strengths than the corresponding deposits of lower carbon content. Manganese had a smaller strengthening effect than at the lower carbon level and lost its strengthening effect with 690 °C PWHT.

Table 9.6 Composition of higher carbon 2.25%Cr–1%Mo (E9018–B3 type) weld metal

Element, wt%

C	Mn	Si	S	P	Cr	Mo
0.092	0.59	0.28	0.008	0.008	2.42	1.02
0.093	0.89	0.27	0.007	0.008	2.42	0.98
0.093	1.24	0.30	0.006	0.008	2.46	1.02
0.094	1.66	0.29	0.006	0.011	2.46	1.01

Note: for typical impurity levels, see Table 3.1.

Table 9.7 Tensile test results from higher carbon 2.25%Cr–0.5%Mo (E9018–B3 type) welds of varying manganese content

Mn, %	State	Yield strength, N/mm^2	Tensile strength, N/mm^2	Elongation, %	Reduction of area, %
0.6	AW	870	950	17	63
1.0		930	1000	16	59
1.4		960	1040	17	62
1.8		970	1050	19	59
0.6	SR	850	1000	17	62
1.0	580 °C	900	1040	18	62
1.4		940	1080	18	59
1.8		960	1120	17	56
0.6	SR	570	670	19	70
1.0	690 °C	560	670	20	71
1.4		600	690	20	69
1.8		580	680	19	69
0.6	STC	560	660	20	66
1.0		560	670	20	71
1.4		550	670	20	69
1.8		500	640	20	67

Note: AW – as-welded; SR – stress relieved; either 2 hr at 580 °C or 20 hr at 690 °C; STC – stress relieved 20 hr at 690 °C and then given step cooling heat treatment.

The relation between Charpy toughness and manganese was complex (Fig. 9.10). As-welded and after PWHT at 580 °C, toughness was very poor (hence only the 28 J temperatures could be determined). As-welded, low manganese was beneficial but after PWHT at 580 °C there was an optimum manganese content at about 1%.

After heat treatment at 690 °C (Fig. 9.11), an optimum of about 0.8%Mn was evident; this fell after step cooling to 0.6% or less. Step cooling itself increased the transition temperature by ~20 °C for weld metal with 0.6%Mn but by as much as ~40°C for weld metals of higher manganese content, as would be expected from Bruscato's formula,[1] which includes the detrimental term (Mn + Si). In practice, a smaller shift in transition temperature can be achieved on step cooling by employing higher purity raw materials than in the present case. Also, notch toughness of standard 2.25%Cr–1%Mo can be considerably enhanced by addition of nitrogen.[2] Modified 2.25%Cr–1%Mo and 3%Cr–1%Mo weld metals, containing vanadium and niobium, however, evidently do not respond to a nitrogen addition.[3]

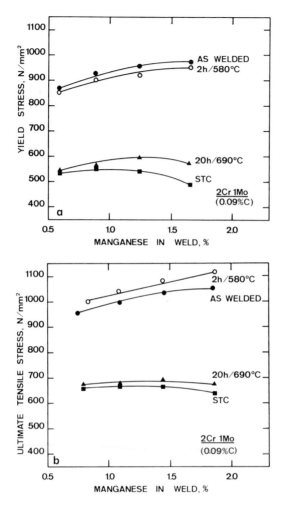

9.9 Effect of manganese on strength of higher carbon 2.25%Cr–1%Mo weld metal after different heat treatments: *a*) Yield strength; *b*) Tensile strength.

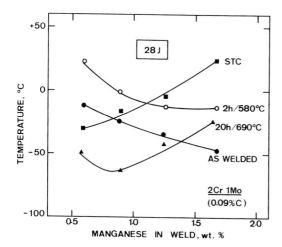

9.10 Effect of manganese on 28 J Charpy transition temperature of higher carbon 2.25%Cr–1%Mo weld metal in different heat treatment conditions.

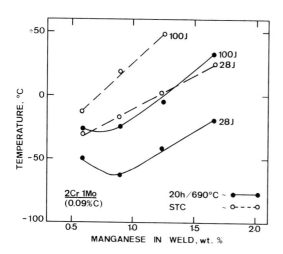

9.11 Effect of manganese on 100 and 28 J Charpy transition temperature of higher carbon 2.25%Cr–1%Mo weld metal after PWHT at 690 °C, showing influence of step cooling.

References

1 Bruscato RM: 'Embrittlement factors for estimation of temper embrittlement in 2¹/₄Cr–1Mo, 3.5Ni–1.75Cr–0.5Mo–0.1V and 3.5Ni steels'. ASTM conf, Miami Fl, Nov 1987.

2 Hojo I and Yamamoto S: 'Improvement of notch toughness by nitrogen addition in $2^1/_4$Cr–1Mo weld metal'. IIW Doc II–1045–85.

3 Nies H, de Giorgi C and Evans GM: 'A new range of welding consumables for modified steels with 2–3%Cr'. Weldex 95, Birmingham, UK.

Part III

High purity weld metal

10

Titanium

To examine effects of microalloying elements on weld metal properties, it was necessary to start from a pure base metal composition. As was seen in Chapter 3, the formulation used for the standard weld metals examined in Parts I and II gave typical contents (in ppm) of approximately 120 V, 55 Ti, 20 Nb, 5 Al and 2 B, together with 50 ppm Mo and ~0.03% each of Cr, Ni and Cu. Most of the last four elements came from the core wire and were thought to be less significant than the five microalloying elements, which were eventually all held below 5 ppm by selection of appropriate raw materials. The same core wire, a typical analysis of which was given in Chapter 2, was used for all parts of the investigation. Weld metal containing 1.4%Mn formed the basis for many of the tests.

Rutile and other minerals

During the course of the investigation, changes in the mineral composition of the electrode coatings were explored, to examine their effects on mechanical properties and weldability. As is shown in Fig. 10.1, the only oxide which had a beneficial effect on toughness was rutile (TiO_2), which is added to electrode coatings to improve their welding behaviour. Other oxides, for example yttria (Y_2O_3), had no detectable effect or, for example alumina, ceria and zirconia (Al_2O_3, CeO_2 and ZrO_2), were detrimental to some degree. Rutile is an impure form of titanium oxide which contains niobium and vanadium among its impurities.

To examine this effect in more detail, balanced formulations were prepared[1] containing up to 10% rutile to give weld metals with a nominal 0.07%C, 1.4%Mn and 0.4%Si. Preparation and testing were identical with earlier tests on C–Mn weld metals. Satisfactory welding behaviour was obtained from all electrodes, and deposits were made free from defects.

Table 10.1 Composition of weld metal deposited from electrodes having varying rutile content

Element, wt%	C	Mn	Si	S	P	Ti	N	O
Rutile in coating, %								
0	0.072	1.41	0.39	0.007	0.005	0.0003	0.007	0.041
1	0.061	1.44	0.33	0.005	0.009	0.0015	0.007	0.038
2	0.076	1.40	0.41	0.005	0.006	0.0027	0.008	0.040
4	0.073	1.38	0.44	0.004	0.006	0.0042	0.007	0.042
6	0.071	1.42	0.44	0.008	0.006	0.0053	0.008	0.040
8	0.073	1.44	0.46	0.008	0.006	0.0069	0.008	0.040
10	0.070	1.40	0.45	0.004	0.007	0.0078	0.008	0.042

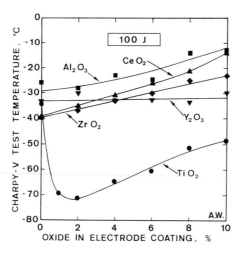

10.1 Effects of different oxides in electrode coating on 100 J Charpy transition temperature.

The compositions of the welds (Table 10.1), show that without rutile the titanium content was less than 5 ppm, and that successive additions increased the weld titanium content to almost 80 ppm with a 10% rutile addition. Other elements were sensibly constant; in particular, rutile had no detectable effect on weld oxygen or nitrogen content.

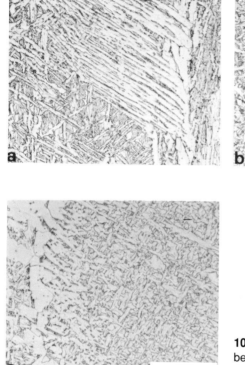

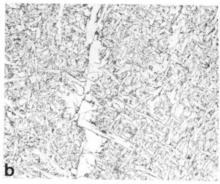

10.2 Microstructure of top as-deposited beads of welds with 1.4%Mn and: *a)* 3 ppm Ti; *b)* 27 ppm Ti; *c)* 78 ppm Ti.

Metallography

Examination of the top, unreheated weld beads showed that an addition of 2% rutile, giving 27 ppm Ti in the weld, altered the microstructure from one of ferrite with aligned second phase and less than 10% acicular ferrite (Fig. 10.2(*a*)), to a predominantly acicular ferrite type (Fig. 10.2(*b*)). Detailed point counting (Fig. 10.3), showed that most acicular ferrite was generated by the first 1% of rutile, giving 15 ppm Ti. The maximum content of acicular ferrite (and the minimum of ferrite with second phase) was obtained at about 25–40 ppm Ti. Increasing the rutile content to 10% (78 ppm Ti) slightly increased the proportion of ferrite with second phase, mainly at the expense of primary ferrite. Besides introducing large quantities of acicular ferrite, the first addition of titanium increased the amount of primary ferrite from about 10 to 20% and removed the ferrite–carbide aggregate. With the highest rutile addition

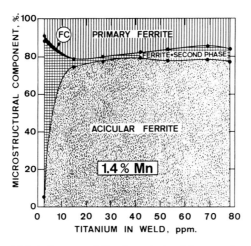

10.3 Influence of titanium derived from rutile additions on microstructural constituents of top, unreheated 1.4%Mn weld beads (FC – ferrite-carbide aggregate).

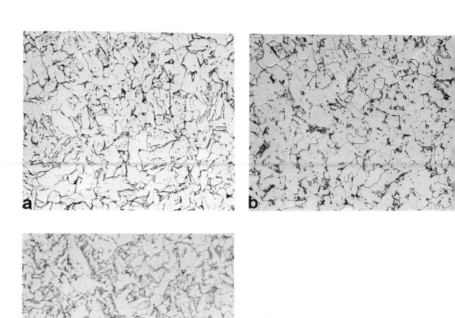

10.4 Effect of rutile additions on microstructure of fine-grained regions: a) No rutile, 3 ppm Ti; b) 2% rutile, 27 ppm Ti; c) 10% rutile, 78 ppm Ti.

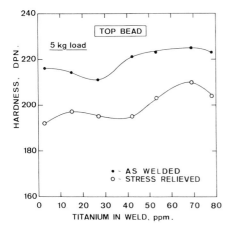

10.5 Effect of titanium from rutile additions on hardness of top (un-reheated) 1.4%Mn weld beads.

(Fig. 10.2(*c*)), the acicular ferrite was of higher aspect ratio than with 1% and some ferrite with unaligned second phase was noted.

The coarse-grained reheated microstructure showed similar changes with titanium content so that, with 2% rutile, the ferrite with second phase was almost absent from the grain interiors. The fine-grained reheated regions (Fig. 10.4) were refined by the addition of titanium; also, the type of second phase was changed. Without rutile, carbide at the ferrite grain boundaries was the predominant constituent. Addition of 2% rutile led to a pearlitic constituent, whilst with 10% rutile the structure was duplex with a high proportion of the M/A constituent. Stress relief precipitated grain boundary carbides in all cases in the fine-grained regions.

Mechanical properties

Figure 10.5 shows that rutile additions initially reduced as-welded hardness slightly and then additions above 2% led to a slightly larger increase. Stress relief reduced hardness by a variable amount.

Table 10.2 and Fig. 10.6 demonstrate the influence of rutile additions on tensile properties. Strength was increased and elongation slightly decreased. The first increase in strength was gradual but as-welded values peaked with a 10% rutile addition. Stress relieving reduced strength levels (particularly yield strength). The strength/Ti plot levelled off at about 50 ppm Ti; it then rose again with the final addition.

Table 10.2 Mechanical properties of 1.4%Mn welds of varying titanium content resulting from rutile additons

State	Ti, ppm	Yield strength, N/mm^2	Tensile strength, N/mm^2	Elongation, %	Reduction of area, %	Charpy transition, °C	
						100 J	28 J
AW	3	460	540	31	77	−34	−64
	15	470	550	31	79	−70	−91
	27	500	550	28	78	−72	−93
	42	500	560	29	79	−65	−92
	53	510	570	30	78	−61	−83
	69	550	600	29	76	−52	−81
	78	540	590	29	78	−49	−70
SR	3	410	500	34	76	−61	−89
	15	440	520	33	77	−79	−105
	27	450	530	34	77	−77	−99
	42	470	540	31	78	−69	−89
	53	460	550	32	77	−61	−80
	69	470	550	29	76	−51	−74
	78	500	580	29	77	−43	−67

Note: AW – as-welded; SR – stress relieved at 580 °C.

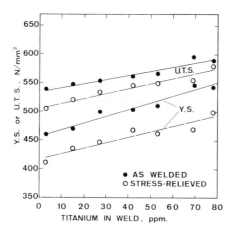

10.6 Effect of titanium from rutile additions on as-welded strength properties of 1.4%Mn multirun deposits.

As-welded Charpy transition temperatures, included in Table 10.2 and plotted in Fig. 10.7, reveal a sudden improvement in toughness as titanium was added. The 100 J temperature fell from about −35 °C to a minimum of almost −80 °C, which occurred close to a 2% rutile

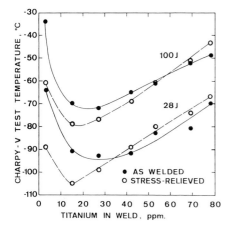

10.7 Effect of titanium from rutile additions on as-welded Charpy transition temperature of 1.4%Mn welds.

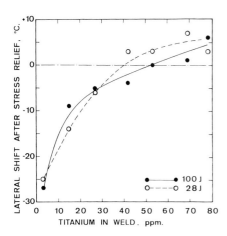

10.8 Change in Charpy transition temperature on stress relieving at 580 °C for 1.4%Mn welds of varying titanium content from rutile additions.

addition (27 ppm Ti). Further additions worsened toughness, although not to the rutile-free levels.

Stress relief improved toughness of the titanium-free deposits the most but rutile was still effective in improving toughness and the minimum Charpy temperature occurred at 15 ppm Ti. However, additions greater than 4% led to transition temperatures higher than for the deposit with no rutile addition. The shifts in transition temperature on stress relief, plotted in Fig. 10.8, show an improvement in Charpy temperature on stress-relief below 40 ppm Ti but a tendency for the transition temperature to increase at higher titanium content.

Titanium metal (Ti)

As rutile is an impure form of titanium oxide, it cannot be used to explore effects of titanium and other microalloying elements in isolation. Several titanium-containing compounds were examined and, although the results are not reported in detail, it can be seen from Fig. 10.9 that all gave an improvement in toughness of the 1.4%Mn weld metal up to a weld metal content of ~20 ppm Ti, followed by a deterioration. The method of addition included metallic titanium added to the electrode coating and this method was used for subsequent tests, although if very small additions were necessary, a relatively pure titanium oxide was used.

In all cases where they were not deliberately added, the weld metals contained <5 ppm each of the microalloying elements Ti, Al, B, Nb and V. However, before the tests on high purity weld metals were undertaken, trials were made on electrodes formulated as before (so-called standard weld metals), except that rutile was absent and tita-

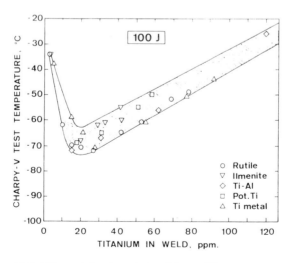

10.9 Effect of titanium derived from different sources on 100 J Charpy transition temperature of pure 1.4%Mn weld metal.

nium was added as metal powder. These electrodes gave weld metals similar to those tested in Part II, whose impurity levels were similar to those given in the footnote to Table 3.1.

Titanium with standard weld metals

These preliminary trials[2] were made with the hitherto optimum manganese content of 1.4%, titanium being added as metal powder to the electrode coating. This avoided minor changes in content of other microalloying elements which occurred when using rutile to vary the titanium level. Welds were tested as-welded only; the compositions of these preliminary welds are given in Table 10.3. The maximum titanium content used was 255 ppm, three times higher than with the highest rutile addition of 10%. The contents of manganese, silicon (except for the last weld) and nitrogen tended to increase as titanium was added; oxygen, however, decreased with titanium content above 30 ppm.

Metallographic examination of the top, un-reheated beads showed similar behaviour to the rutile series below about 50 ppm Ti, as shown by a comparison of Fig. 10.10 and 10.3. As little as 16 ppm Ti gave a predominantly acicular ferrite microstructure with no ferrite-carbide aggregate; the maximum of ~70% acicular ferrite occurred at about 30 ppm Ti. Acicular ferrite then fell to a minimum at ~80 ppm Ti and rose again to nearly 80% at the highest titanium level of 255 ppm. Primary ferrite increased as acicular ferrite reached

Table 10.3 Composition of standard 1.4%Mn weld metal having varying titanium content

Element, wt%	C	Mn	Si	S	P	Ti	N	O
Electrode code								
P	0.074	1.45	0.33	0.005	0.005	0.0006	0.009	0.041
—	0.073	1.49	0.34	0.004	0.007	0.0016	0.008	0.040
—	0.071	1.46	0.36	0.004	0.006	0.0022	0.008	0.040
Q	0.061	1.36	0.33	0.005	0.006	0.0028	0.008	0.042
—	0.070	1.48	0.32	0.005	0.006	0.0055	0.008	0.037
—	0.068	1.38	0.32	0.004	0.005	0.0077	0.007	0.038
R	0.068	1.51	0.35	0.006	0.006	0.0100	0.009	0.035
—	0.065	1.53	0.39	0.005	0.006	0.0140	0.009	0.035
S	0.070	1.64	0.48	0.005	0.006	0.0210	0.010	0.029
—	0.075	1.75	0.29	0.006	0.005	0.0255	0.010	0.027

Note: all welds had impurity levels similar to those given in Table 3.1.

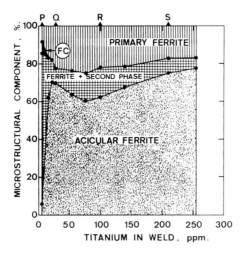

10.10 Influence of weld titanium content on as-deposited microstructure of 1.4%Mn welds with titanium added as Ti powder but 'standard' impurity levels (FC – ferrite-carbide aggregate).

its first maximum (as was the case in the rutile series) but it then remained roughly constant, so that subsequent microstructural changes were accompanied by changes in the proportion of ferrite with second phase.

Changes in coarse-grained reheated weld metal microstructure resembled those in as-deposited regions. Ferrite with second phase had disappeared from the grain interiors at 30 ppm Ti but reappeared with 100 ppm.

As with the rutile series, the fine grained weld metal was refined by titanium. Small additions of titanium also replaced grain boundary films of cementite by degenerate pearlite (B/P or bainite/pearlite), whilst more titanium (100 ppm) altered this to the M/A constituent and ultimately a duplex structure.

Because of unintended variations in composition, changes in hardness and tensile properties with titanium were irregular and are not reported here but are detailed in Ref. 2.

Charpy impact tests, summarised in Fig. 10.11, showed a progressive improvement in as-welded toughness from 100 J at about −35 to about −70 °C, as the weld metal titanium content increased from 6 to 28 ppm; this was followed by a smaller reduction in toughness up to about 140 ppm and then a slight improvement. However, these results are inevitably skewed by the unintended variations in composition noted in Table 10.3.

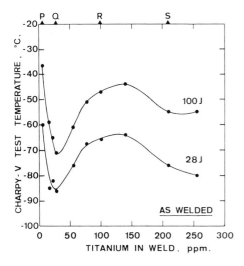

10.11 Effect of titanium (from titanium powder) on as-welded Charpy transition temperature of standard 1.4%Mn weld metal.

Titanium additions to high purity weld metal (Ti)

Trials[3] were made on 1.4%Mn weld metals with titanium contents between 1 and 550 ppm with all other microalloying elements each below 5 ppm (Table 10.4). Apart from an increase in silicon and a reduction in oxygen as the titanium content was increased, the compositions were well balanced.

Microstructural changes were similar to those described for the two earlier series, except that without titanium the proportion of acicular ferrite was appreciably below 10%, instead of just above in the previous two, less pure series.

Mechanical properties are given in Table 10.5. Strength increased as titanium was increased, although some of this may have been a result of accompanying changes in silicon and possibly oxygen contents.

Transition temperatures for 100 J Charpy energy are plotted in Fig. 10.12 and show a similar effect of titanium to the previous two series (Fig. 10.7 and 10.11), confirming the minimum in toughness (maximum transition temperature) at about 100 ppm Ti. In the absence of added titanium, the purer weld metal has, however, given an even higher transition temperature of approximately −15 °C compared with −35 °C in the two less pure earlier series, although the lowest transition temperature was similar. This higher transition

Table 10.4 Composition of high purity 1.4%Mn weld metal having varying titanium content

Element, wt%

C	Mn	Si	S	P	Ti	N	O
0.074	1.40	0.25	0.008	0.007	0.0001	0.008	0.048
0.077	1.46	0.27	0.008	0.007	0.0028	0.008	0.046
0.073	1.41	0.26	0.007	0.011	0.009	0.008	0.039
0.074	1.45	0.30	0.007	0.009	0.015	0.008	0.034
0.072	1.44	0.35	0.006	0.012	0.020	0.007	0.031
0.072	1.50	0.38	0.006	0.010	0.026	0.008	0.029
0.069	1.45	0.40	0.006	0.008	0.033	0.008	0.028
0.069	1.47	0.45	0.005	0.006	0.041	0.008	0.028
0.069	1.49	0.43	0.004	0.005	0.046	0.008	0.029
0.070	1.50	0.45	0.005	0.006	0.055	0.008	0.029

Note: all welds contained <5 ppm each of Al, B, Nb and V, ~0.03%Cr, Ni and Cu, 0.005%Mo.

Table 10.5 Tensile test results for high purity 1.4%Mn welds having varying titanium content

State	Ti, ppm	Yield strength, N/mm^2	Tensile strength, N/mm^2	Elongation, %	Reduction of area, %	Charpy transition, °C	
						100 J	28 J
AW	1	440	530	28	78	−14	−42
	28	470	540	25	77	−68	−88
	90	480	540	24	79	−50	−72
	150	480	540	28	80	−43	−59
	200	500	560	27	79	−53	−76
	260	510	580	26	79	−56	−72
	330	510	560	26	81	−59	−73
	410	500	580	26	80	−61	−77
	460	540	590	26	80	−60	−78
	550	530	600	25	82	−60	−77
SR	1	440	530	28	78	−65	−93
	28	470	540	25	77	−70	−90
	150	480	540	28	80	−56	−74
	260	510	580	26	79	−67	−83
	410	500	580	26	80	−69	−89
	550	530	600	25	82	−63	−79

Note: AW – as-welded; SR – stress relieved at 580 °C.

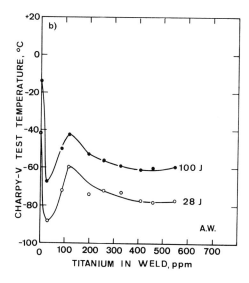

10.12 Effect of titanium (from titanium powder) on as-welded Charpy transition temperature of high purity 1.4%Mn weld metal.

temperature in pure Ti-free weld is consistent with the lower acicular ferrite content. After stress relief, the changes in toughness with titanium additions were less dramatic because of the much improved toughness of the titanium-free deposit after stress relief. Nevertheless, apart from the much smaller initial improvement, the changes in toughness reflected those in the as-welded condition. Unlike the rutile series (Fig. 10.8), where the weld of highest titanium content (~80 ppm) had a slightly higher transition temperature after PWHT, the stress-relieved welds were *always* tougher.

Titanium and manganese (Mn–Ti)

Because the foregoing tests had emphasised the importance of small additions of titanium and other microalloying elements in controlling microstructure and toughness, a further series was carried out with the full range of manganese content,[4] for which the electrodes were made of ingredients giving <5 ppm of each microalloying element in the weld metal. For future reference, compositions similar to those previously referred to as A, B, C and D but containing no titanium have been coded as A_0, B_0, C_0 and D_0; the original designations are used for compositions with ~30 ppm Ti. Tests were performed both as-welded and after PWHT for 2 hours at 580 °C.

Table 10.6 Composition of weld metal having varying titanium and manganese content

Element, wt%	C	Mn	Si	S	P	Ti	N	O
Electrode code								
A_0	0.074	0.63	0.35	0.009	0.007	0.0008	0.007	0.047
A	0.074	0.65	0.35	0.005	0.004	0.0036	0.007	0.043
	0.075	0.66	0.35	0.009	0.007	0.0096	0.007	0.040
	0.076	0.67	0.38	0.009	0.007	0.0160	0.007	0.036
	0.078	0.71	0.44	0.006	0.008	0.0225	0.008	0.031
B_0	0.077	1.00	0.30	0.008	0.007	0.0007	0.008	0.046
B	0.076	1.01	0.35	0.008	0.005	0.0041	0.007	0.040
	0.075	1.03	0.31	0.008	0.007	0.0092	0.008	0.039
	0.075	1.05	0.35	0.008	0.007	0.0140	0.008	0.036
	0.078	1.10	0.43	0.007	0.005	0.0190	0.007	0.033
C_0	0.075	1.31	0.30	0.008	0.006	0.0005	0.008	0.044
C	0.077	1.33	0.32	0.007	0.005	0.0036	0.008	0.042
	0.073	1.32	0.30	0.008	0.007	0.0075	0.008	0.040
	0.079	1.35	0.33	0.008	0.007	0.0140	0.008	0.038
	0.080	1.37	0.39	0.007	0.006	0.0215	0.009	0.032
D_0	0.077	1.79	0.28	0.007	0.006	0.0007	0.008	0.044
D	0.078	1.75	0.31	0.007	0.006	0.0036	0.009	0.042
	0.077	1.76	0.29	0.007	0.007	0.0105	0.008	0.040
	0.075	1.83	0.32	0.006	0.007	0.0170	0.008	0.035
	0.079	1.82	0.37	0.006	0.007	0.0255	0.008	0.032

Notes: the values are the average for as-welded and stress-relieved deposits.
All welds contained <0.005%Nb, V, Al, B, ~0.03%Cr, Ni and Cu, 0.005%Mo.

Composition

The compositions of the welds (Table 10.6), were better balanced than in the preliminary series (Table 10.3), although oxygen still fell with increasing titanium whilst silicon and (to a smaller degree) manganese rose slightly in each series. The relationship between titanium and oxygen content is further explored in the next chapter and in Chapter 14.

Metallography

Examination of the top, un-reheated beads of deposits (Fig. 10.13), showed that in the series containing 1.4 and 1.8%Mn, a small addition of ~30 ppm Ti in the weld metal increased the proportion of acicular ferrite from about 10 to about 70% (Fig. 10.13(*c,d*)). The tests on the rutile and preliminary series had shown the increase to be a progressive one. Titanium contents between 50 and 100 ppm gave a

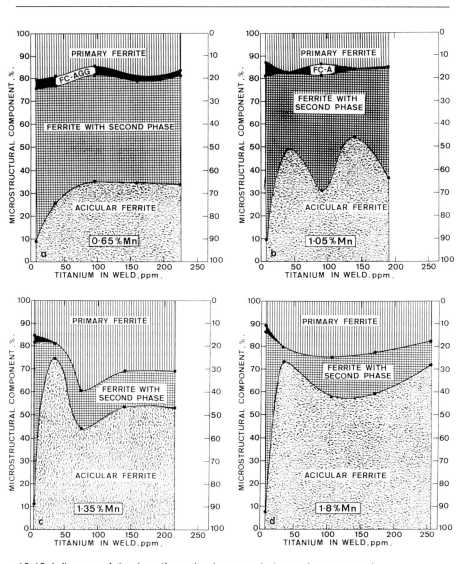

10.13 Influence of titanium (from titanium powder) on microstructural constituents of top, unreheated beads of high purity welds with: *a*) 0.65%Mn; *b*) 1.05%Mn; *c*) 1.35%Mn; *d*) 1.8%Mn (FC-A or FC-AGG – ferrite-carbide aggregate).

decrease in the proportion of acicular ferrite, although this was largely recovered by larger additions.

The weld metals of lowest manganese content (0.6%), however, showed a gradual increase in acicular ferrite with increasing titanium

Table 10.7 Tensile test results from welds having varying titanium and manganese content: *a*) As-welded

Mn, %	Ti, ppm	Yield strength, N/mm²	Tensile strenght, N/mm²	Elongation, %	Reduction of area, %	Charpy transition, °C 100 J	28 J
0.6, A₀	8	400	490	32	76	−46	−70
A	36	400	490	33	78	−45	−70
	96	400	490	30	78	−40	−59
	160	410	490	31	79	−36	−58
	225	440	520	31	77	−40	−64
1.0, B₀	7	410	500	31	77	−56	−83
B	41	440	520	31	77	−60	−84
	92	430	520	32	79	−53	−73
	140	430	520	31	79	−45	−63
	190	460	540	29	79	−50	−72
1.4, C₀	5	420	520	26	77	−29	−49
C	36	460	540	27	76	−68	−89
	75	450	520	29	79	−52	−67
	140	460	540	29	79	−44	−67
	215	480	560	28	81	−54	−74
1.8, D₀	7	460	570	29	76	−5	−40
D	36	500	570	29	77	−62	−90
	105	480	580	26	78	−54	−82
	170	510	590	27	80	−52	−82
	255	540	610	25	76	−53	−79

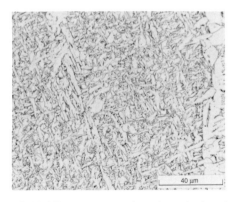

10.14 Microstructure of as-deposited region of 1%Mn weld with 140 ppm Ti.

Table 10.7 (cont.) b) Stress-relieved at 580 °C

Mn %	Ti, ppm	Yield strength, N/mm^2	Tensile strength, N/mm^2	Elongation, %	Reduction of area, %	Charpy transition, °C	
						100 J	28 J
0.6, A$_0$	7	330	450	30	79	−53	−79
A	36	330	450	36	80	−55	−72
	96	350	470	32	80	−54	−68
	160	360	460	30	80	−56	−66
	225	380	480	32	78	−51	−71
1.0, B$_0$	7	370	480	30	79	−66	−90
B	41	350	480	29	80	−80	−93
	92	380	470	32	80	−56	−75
	140	400	490	32	81	−54	−75
	190	400	490	32	81	−56	−77
1.4, C$_0$	5	360	480	30	79	−72	−100
C	36	370	500	31	76	−81	−97
	75	420	510	33	81	−61	−83
	140	400	510	27	80	−59	−81
	215	430	540	28	78	−70	−86
1.8, D$_0$	7	380	500	30	79	−49	−90
D	36	410	520	29	78	−71	−99
	105	430	530	27	79	−58	−90
	170	470	560	28	78	−58	−85
	255	470	580	29	78	−61	−84

(Fig. 10.13(*a*)), reaching a plateau of about 35% acicular ferrite at 100 ppm Ti. The 1%Mn weld metals (Fig. 10.13(*b*)), although exhibiting a similar sharp increase in acicular ferrite to the deposits of higher manganese content, showed a sharper dip and a sharper rise to a higher peak at 140 ppm Ti. The behaviour of the two lower manganese series may, however, have also been influenced by the slight increase in manganese through each series because, even if conditions are favourable for nucleation of acicular ferrite, sufficient alloying (in this case with manganese) is necessary to ensure its formation.

In all series, acicular ferrite was produced chiefly at the expense of ferrite with aligned second phase, although amounts of primary ferrite and ferrite/carbide aggregate also changed, the latter being eliminated in the two welds of higher manganese content with 35 ppm Ti, as was the case with the earlier series.

The influence of the small additions of titanium has already been described; with higher titanium content, the acicular ferrite tended to have a greater aspect ratio (Fig. 10.14). The microstructure of the coarse-grained reheated regions was generally similar to the as-deposited weld metal.

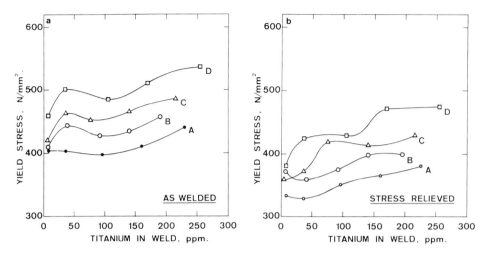

10.15 Influence of weld titanium and manganese content on yield strength: *a)* As-welded; *b)* Stress relieved at 580 °C.

Except for the welds of lowest manganese content, titanium refined the fine-grained reheated regions (as in the preliminary series) and progressively altered their structures. Small titanium additions replaced grain boundary carbides by degenerate pearlite and increasing amounts by the M/A (martensite/austenite) constituent. Stress relief led to globularisation of the carbides in the M/A constituent; this was less marked in welds of lower manganese content, where degenerate pearlite was more resistant to globularisation.

A description of the influence of titanium on the structure and composition of non-metallic inclusions is given in Chapter 17.

Mechanical properties

Hardness values of the top, unreheated beads changed by less than ±10 HV5 in each of the main series welds, so that the influence of titanium was minor.

Tensile test results for the series are included in Table 10.7 and the yield strength values are plotted against titanium content in Fig. 10.15. There was a tendency for strength to increase with titanium, although some of this may have been a result of the small increases in manganese and silicon, particularly for the welds of lowest manganese content. Otherwise the rises and falls in yield strength tended to reflect the changes in acicular ferrite content.

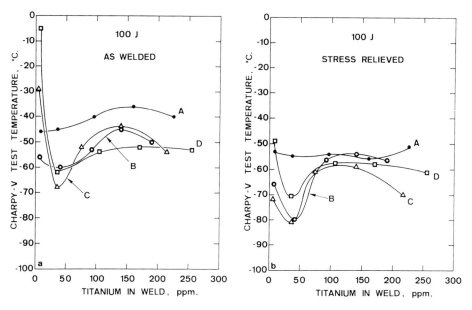

10.16 Influence of weld titanium and manganese content on 100 J Charpy transition temperature: *a*) As-welded; *b*) Stress relieved at 580 °C.

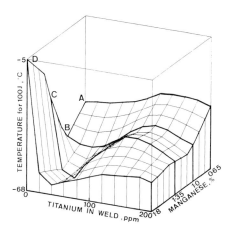

10.17 Model of as-welded 100 J Charpy transition temperature for Mn–Ti weld metals.

Stress relief gave an overall reduction in strength, particularly at high titanium levels, which can be attributed to the M/A constituent being more likely to soften on tempering than the degenerate pearlite found in the low titanium deposits.

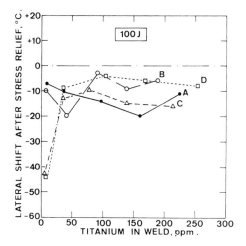

10.18 Change in 100 J Charpy transition temperature on stress relief for deposits of varying titanium and manganese content.

As-welded Charpy tests, summarised in Fig. 10.16(*a*), showed that the dramatic effect of small additions (~40 ppm) of titanium on as-welded toughness was seen only with the 1.4 and 1.8%Mn deposits (weld series C and D). The 1.0%Mn weld B exhibited only a slight improvement and the 0.6%Mn weld A none at all. Titanium contents above about 40 ppm led to a small worsening in toughness of all deposits. This was, in all except the deposit of highest manganese content, succeeded by a small improvement. The lowest transition temperature of nearly −70 °C occurred with the 1.4%Mn deposit with 36 ppm Ti, thus confirming that the optimum manganese level was unchanged by use of purer weld metals. However, a three-dimensional model (Fig. 10.17), revealed that 1.4%Mn was only optimum for a restricted range of titanium contents around 30 ppm and moved to higher manganese levels (1.7–1.8%) as titanium was increased.

In the stress-relieved condition (Fig. 10.16(*b*)), the small titanium addition gave a noticeable improvement in toughness with all welds except those made with electrode A of the lowest (0.6%) manganese content. Stress relief gave an improvement in transition temperature (Fig. 10.18) of about 10 °C with all except the nominally titanium-free deposits at the two highest manganese contents (Fig. 10.16), where the improvement was even more marked (40–50 °C). It is likely that the low concentrations of other microalloying elements avoided the increases in transition temperature on stress relief found with less pure C–Mn weld metals described in Chapter 3 and earlier in this chapter.

Simulation tests

To study the changes in properties of different regions of typical weld deposits brought about by the small addition of titanium needed to alter the microstructure to one of acicular ferrite, specimens of 1.4%Mn weld metal were treated using a Gleeble weld simulator and

Table 10.8 Chemical analysis of 1.4%Mn all-weld metal deposits for simulation tests

Element, wt%	Electrode O	Electrode W
C	0.074	0.077
Mn	1.40	1.46
Si	0.25	0.27
S	0.008	0.008
P	0.007	0.007
Ti	0.0001	0.0028
B	0.0001	0.0003
Al	0.0006	0.0005
N	0.008	0.008
O	0.048	0.046

Note: both welds contained <0.0005%Nb, V, ~0.03%Cr, Ni and Cu, 0.005%Mo.

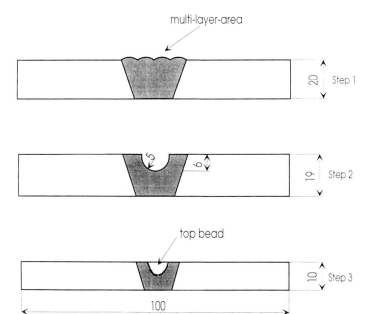

10.19 Preparation of samples for simulation (dimensions in mm).

V-notch

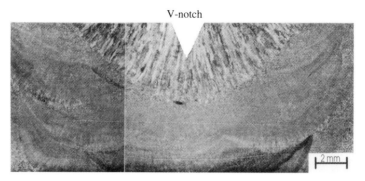

10.20 Charpy notching of simulated samples, macro-etched specimen.

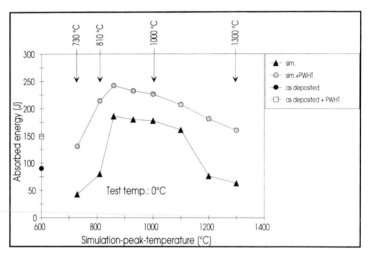

10.21 Effect of simulation peak temperature on Charpy toughness at 0 °C of Ti-free weld metal O; as-welded and simple stress relieved results shown at left.

tested.[5] The electrodes selected gave all-weld analyses as shown in Table 10.8, the weld from electrode O containing 1 ppm Ti and that from electrode W 28 ppm; otherwise the compositions were as close as possible to each other.

After preparing a typical all-weld deposit, the weld was grooved as in Fig. 10.19 and a single weld run deposited at 3.2 kJ/mm (220 A, 24 V). After machining to size (Step 3), specimens were given a weld thermal simulation, in some cases followed by stress relief heat treat-

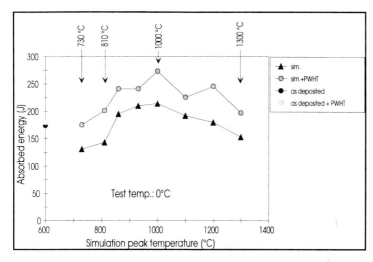

10.22 Effect of simulation peak temperature on Charpy toughness at 0 °C of weld metal W with Ti; as-welded and simple stress relieved results shown at left.

ment of 2 hours at 580 °C. Standard Charpy specimens were then prepared so that the notch lay in as-deposited weld metal (Fig. 10.20) and tested at 0 °C or, for selected simulations, over a wider range of temperature. For comparison, similar specimens were tested without simulation in the as-welded or stress relieved conditions.

From dilatometric studies, which had shown the Ac_1 and Ac_3 temperatures to be 750 and 880 °C, respectively, peak simulation temperatures of 730, 810, 860, 930, 1000, 1100, 1200 and 1300 °C were selected and the time/temperature cycles were chosen to represent different regions of welds deposited at 1 kJ/mm. Metallographic examination was carried out by light microscopy.

Without titanium, subcritical, low intercritical or high supercritical temperature simulation at 730, 810, 1200 and 1300 °C gave Charpy energies similar to or slightly below the as-welded value (Fig. 10.21). High intercritical and low supercritical simulation at 860–1100 °C resulted in improved toughness, values falling as the peak temperature was increased. After stress relief, toughness improved, except with the 730 °C simulation, although the values again tended to fall as the peak temperature was increased.

Toughness levels at 0 °C were generally higher in the weld metal containing titanium (Fig. 10.22), although as-welded, the same pattern emerged. Peak temperatures of 730, 1200 and 1300 °C gave

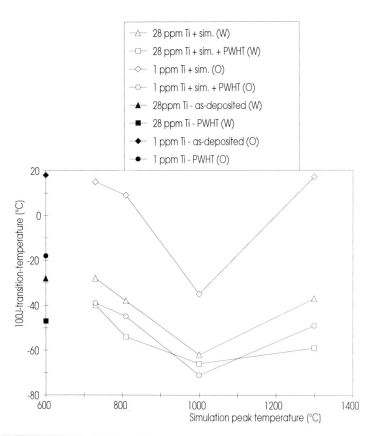

10.23 100 J Charpy transition temperature for deposits with and without titanium after different peak simulation temperatures; as-welded and simple stress relieved results shown at left.

distinctly lower values than as-welded, whilst 860–1100 °C were only slightly improved. Although PWHT gave no improvement to the welded samples, peak temperatures followed by stress relief gave a marked improvement with all except the highest and lowest peak temperatures.

Sufficient specimens were simulated to peak temperatures of 730, 810, 1000 and 1300 °C (i.e. sub-critical, low intercritical, mid fine-grained supercritical and high supercritical temperatures) to enable 100 J Charpy temperatures to be determined (Fig. 10.23). Although the simulated Ti-free specimens all gave lower transition temperatures than the corresponding as-welded value, it should be noted that the

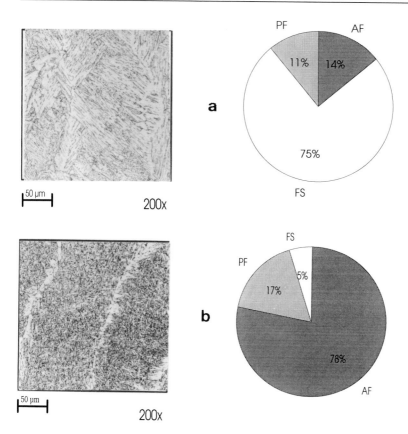

10.24 As-deposited microstructure of 1.4%Mn weld metal: *a*) With 1 ppm Ti; *b*) With 28 ppm Ti.

as-welded result was from a 3.2 kJ/mm deposit whilst the simulated results represented 1 kJ/mm welds (the higher heat input would be expected to result in inferior toughness and the same comments are applicable to the Ti-bearing results). However, only the 1000 °C specimens showed appreciably greater toughness than the as-welded result and even that was of similar toughness to most of the specimens from the weld containing titanium. The stress-relieved Ti-free and the Ti-bearing specimens all gave similar results, the toughness generally improving as the peak simulation temperature increased to 1000 °C and then declining at 1300 °C.

However, although the simulated plus PWHT specimens gave similar results for Ti-bearing and Ti-free deposits, the stress-relieved Ti-free weld itself gave a 100 J transition temperature as much as

Table 10.9 Microstructure of 1.4%Mn weld metal, with and without titanium, after simulation

Ti, ppm	Peak temperature, °C	Microstructural content, vol%			
		Primary ferrite	Acicular ferrite	Ferrite with aligned second phase	Ferrite/carbide aggregate
1	As-welded	12	13	74	<1
	730	10	11	78	1
	810	7	14	78	1
	1000	-------------- Too fine to classify --------------			
	1300	4	8	88	0
28	As-welded	16	79	4	0
	730	18	78	4	0
	810	20	78	2	0
	1000	-------------- Too fine to classify --------------			
	1300	38	58	4	0

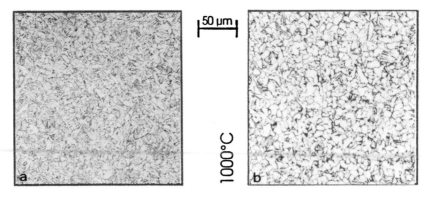

10.25 Fine-grained microstructure of 1.4%Mn weld metal after simulation to 1000 °C: *a*) With 1 ppm Ti; *b*) With 28 ppm Ti.

30 °C higher than the corresponding stress-relieved weld containing titanium. This result must put some doubt on the suitability of the simulation technique for evaluating all aspects of weldment toughness behaviour.

Metallographic examination (Fig. 10.24) showed that the Ti-free weld metal was predominantly ferrite with aligned second phase whereas with 28 ppm Ti acicular ferrite predominated. Reheating to 730 and 810 °C changed the proportions of the constituents only a little (Table 10.9). At the peak simulation temperature of 1000 °C

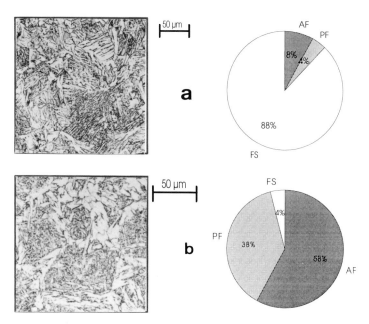

10.26 Coarse-grained microstructure of 1.4%Mn weld metal after simulation to 1300 °C: *a*) With 1 ppm Ti; *b*) With 28 ppm Ti.

(which gave the highest toughness) both welds exhibited uniform refined microstructures (Fig. 10.25), which could not be classified by the IIW scheme (which is intended for as-deposited and similar microstructures). After a highest peak temperature of 1300 °C, the microstructures (Fig. 10.26) were similar to the coarse-grained reheated weld metal regions, the difference being that, with titanium the amount of primary ferrite was noticeably greater than as-welded, whereas without Ti, both primary and acicular ferrite were reduced. This difference corresponds to the relatively inferior toughness of the Ti-free deposit in this condition.

Discussion

The change in microstructure brought about by the small change in micro-alloying elements was remarkable and is summarised in Table 10.9. The purer weld metal appeared to lack an ingredient capable of nucleating acicular ferrite – namely titanium in some form.

The failure to nucleate acicular ferrite in pure C–Mn weld metals has two major effects. The better known is inferior toughness when acicular ferrite is replaced by ferrite with aligned second phase. This is a result of fine (ferrite) grained acicular ferrite being replaced in the as-deposited and coarse-grained reheated microstructures by the much coarser and, therefore, less tough constituent. The second effect is that the strengthening effect of manganese is much reduced – indeed lost in the stress-relieved condition. Again, this is a result of manganese promoting acicular ferrite in suitable conditions and thus refining the overall grain size.

General discussion

Although the profound effect of as little as 20 ppm Ti on the microstructure and toughness of weld metal has been seen also in submerged-arc and MIG welds (Ref. 8, Ch. 3 and Ref. 6 and 7, this chapter), the present work shows that the effect was also critically dependent on an adequate manganese content and is much reduced in welds after stress relief. Furthermore, in both types of submerged-arc weld reported previously, aluminium was present in the weld metal from the fluxes used and, in one case,[6] niobium was deliberately added to the welds via the parent steel, whilst in the other no attempt was made to avoid contamination by small amounts of micro-alloying elements.

The results presented here provide a clear and unambiguous demonstration of the influences of titanium on the microstructure and as-welded toughness of C–Mn weld metal containing no significant amounts of other microalloying elements. The absence of such elements was, indeed, found to worsen both the microstructure and toughness of C–Mn weld metal without titanium (C_0). Without any microalloying elements, microstructures contained less than 10% acicular ferrite and gave 100 J at about −15 °C. With traces of such elements (typically 120 ppm V, 20 ppm Nb, 5 ppm Al, 3 ppm Ti and 2 ppm B), deposits contained more than 10% acicular ferrite with 100 J Charpy temperatures of about −35 °C. With 30 ppm Ti, the microstructures were predominantly of acicular ferrite with 100 J temperatures of about −70 °C, more or less regardless of traces of other microalloying elements. Effects of other microalloying elements are studied in Chapter 13 and other combinations of them in Chapter 14.

After stress-relief, the effect of titanium on toughness was less spectacular although perceptible. However, the absence of microalloying elements other than titanium led to Charpy levels falling on stress relief, coupled with slightly lower yield strength values (these were also noted in the as-welded condition).

It must be emphasised that the influence of titanium on as-deposited and coarse-grained microstructures is reinforced by its effects on fine-grained regions. In all series, titanium produced a slight refinement of these regions but – more significantly – it replaced grain boundary carbide films with degenerate pearlite. The importance of removing such films is highlighted by the excellent toughness of Ti-free weld metal after stress relief – a result of the grain boundary carbides having been effectively removed by being globularised during PWHT.

Titanium is only effective provided there is sufficient alloying with manganese (possibly aided by other major alloying elements). With only 0.6%Mn, the weld metal was insufficiently alloyed for acicular ferrite to be developed and 30 ppm Ti hardly altered the 100 J Charpy temperature and more was slightly adverse (Fig. 10.16(a)).

Effects of minute amounts of vanadium ($\sim$120 ppm) and niobium ($\sim$20 ppm) present in the less pure weld metals discussed in Part I are probably of major importance. In the pure weld metals in the present chapter, stress relief heat treatment invariably improved toughness and lowered transition temperatures. In the less pure rutile series (Fig. 10.8), stress relief improved toughness only when the titanium content was low, and in the less pure C–Mn weld metals in Chapter 5, stress relief also increased transition temperatures when carbon and manganese where high. These results point to vanadium (possibly helped by niobium and other factors such as carbon content) impairing toughness after stress relief. The synergistic effect of vanadium and niobium on HAZ toughness after PWHT has already been noted in Ref. 8; effects of larger quantities of the two elements (added singly) on weld metal toughness are discussed in Chapter 13.

References

1 Evans GM: 'The effect of rutile in the coating of a basic low hydrogen MMA electrode'. IIW Asian Pacific Welding Congress, Auckland, New Zealand, Feb 1996, IIW DocIIA–990–96.
2 Evans GM: 'Effect of titanium on the microstructure and properties of C–Mn all-weld-metal deposits'. *Weld J* 1992 **71**(12) 447s–54s; *Welding Res Abroad* 1992 **38**(8/9) 13–21; *Schweissmitteilungen* 1991 **49**(125) 22–33.
3 Evans GM: 'Effect of micro-alloying elements in C–Mn steel weld metal'. *Welding in the World* 1993 **31**(1) 12–19; *Schweissmitteilungen* 1993 **51**(129) 16–26.
4 Evans GM: 'Effect of Ti in Mn-containing MMA weld deposits'. *Weld J* 1992 **71**(12) 447s–54s; *Schweissmitteilungen* 1991 **49**(128) 19–34.

5 Cerjak H, Letofsky E, Pitoiset X, Seiringer A and Evans GM: 'The influence of the microstructure on the toughness of C–Mn multi-run-weld metal' IIW Doc IX–1814–95.

6 Terashima H and Hart PHM: 'Effect of aluminium in C–Mn–Nb steel submerged-arc welds'. *Weld J* 1984 **63** 173s–183s.

7 Hart PHM and Hutt GA: 'An investigation into the factors influencing mechanised MIG weld metal toughness'. Proc conf 'Welding and performance of Pipelines', London, 1986, TWI, Paper 71.

8 Rothwell AB: 'HAZ toughness of welded joints in microalloyed steels: Part 1 – Results of instrumented impact tests'. CANMET Report No. 79–6, Jan 1979.

11

Combinations with oxygen

It became apparent during the investigation that the profound influence of titanium on weld metal microstructure and toughness was connected with inclusions containing titanium – presumably as an oxide (Ref. 2, Ch. 1). It was important, therefore, to examine effects of both oxygen and nitrogen on weld properties. Nitrogen is dealt with in the next chapter and oxygen is the subject of this. Oxygen was studied in connection with the three elements titanium, manganese and aluminium. The first and last of these have not been reported previously.

Titanium and oxygen (Ti–O)

Electrodes were made to give high purity 1.4%Mn weld metals with titanium contents ranging from 2 to 110 ppm (i.e. concentrating on the lower end of the range examined in Chapter 10), in which the weld oxygen content was reduced in four stages from its 'normal' level of about 450 ppm to 250 ppm by additions of metallic magnesium to the coating. The compositions of the resultant welds are given in Table 11.1. Not only did magnesium reduce the oxygen content but also, to a lesser extent, it reduced sulphur and nitrogen. Formulations were as well balanced as possible; nevertheless, silicon tended to increase as titanium increased and oxygen decreased; the change due to titanium was reduced as oxygen was decreased. Otherwise compositions were reasonably satisfactory, except that the manganese contents were a little higher than intended in the 280 ppm O series and a little lower with 350 ppm O.

Tensile and Charpy tests were carried out for both as-welded and stress-relieved conditions. The results are given in Table 11.2.

Strength tended to increase as titanium was increased and oxygen reduced (Fig. 11.1), although the changes were small and the results

Table 11.1 Composition of 1.4%Mn weld metal with varying titanium and oxygen content

Element, wt%	C	Mn	Si	S	P	Ti	N	O
O, ppm								
240	0.075	1.45	0.38	0.003	0.009	0.0003	0.005	0.025
	0.077	1.48	0.39	0.003	0.009	0.0016	0.006	0.023
	0.080	1.46	0.38	0.003	0.009	0.0027	0.006	0.025
	0.084	1.44	0.38	0.003	0.009	0.0035	0.005	0.024
	0.084	1.43	0.36	0.003	0.009	0.0053	0.005	0.025
	0.086	1.49	0.40	0.003	0.009	0.0097	0.005	0.025
280	0.080	1.63	0.35	0.004	0.010	0.0005	0.006	0.028
	0.080	1.65	0.34	0.004	0.010	0.0012	0.006	0.028
	0.087	1.56	0.34	0.004	0.010	0.0022	0.006	0.028
	0.080	1.59	0.35	0.004	0.010	0.0031	0.006	0.028
	0.083	1.56	0.37	0.004	0.009	0.0057	0.007	0.028
	0.086	1.56	0.39	0.004	0.011	0.0090	0.007	0.028
350	0.079	1.36	0.33	0.005	0.011	0.0005	0.008	0.036
	0.081	1.38	0.34	0.006	0.012	0.0015	0.008	0.035
	0.080	1.52	0.38	0.006	0.013	0.0027	0.008	0.034
	0.085	1.56	0.40	0.006	0.014	0.0050	0.009	0.033
	0.083	1.45	0.35	0.006	0.013	0.0069	0.008	0.035
	0.084	1.49	0.44	0.006	0.011	0.0110	0.008	0.034
410	0.079	1.29	0.29	0.007	0.008	0.0004	0.009	0.044
	0.079	1.33	0.31	0.006	0.007	0.0021	0.008	0.042
	0.084	1.37	0.35	0.007	0.008	0.0035	0.010	0.041
	0.082	1.30	0.35	0.007	0.007	0.0053	0.010	0.041
	0.076	1.33	0.40	0.008	0.008	0.0069	0.009	0.041
	0.079	1.34	0.42	0.007	0.009	0.0110	0.010	0.038

Notes: the values are the average for as-welded and stress-relieved deposits.
All welds contained <5 ppm Al, B, Nb, V, ~0.03%Cr, Ni and Cu, 0.005%Mo.

somewhat scattered. The effect of titanium was more marked as-welded and that of oxygen in the stress-relieved condition.

The results of Charpy toughness tests, in terms of transition temperatures (Fig. 11.2), were somewhat less scattered and, in all cases, oxygen did not fundamentally alter the effect of titanium discussed in Chapter 10. However, for both as-welded and stress-relieved deposits, reducing oxygen reduced Charpy temperature, particularly near the point where titanium had its greatest effect on toughness, i.e. around 35 ppm. There appeared to be an optimum oxygen content around 280 ppm, so that the further reduction to 240 ppm gave little or no further benefit.

A second effect of oxygen was that reducing oxygen tended to reduce the amount of titanium required to achieve its maximum effect. This effect was not very pronounced and, for this reason, the

Table 11.2 Mechanical properties of 1.4%Mn welds with varying titanium and oxygen content: *a*) As-welded

O, ppm	Ti, ppm	Yield strength N/mm²	Tensile strength, N/mm²	Elongation, %	Reduction of area, %	Charpy transition, °C	
						100 J	28 J
240	3	470	540	30	80	−37	−55
	16	460	550	27	79	−82	−98
	27	450	550	29	81	−81	−96
	35	470	560	28	82	−79	−94
	53	480	570	26	82	−71	−86
	97	500	590	25	82	−61	−82
280	5	450	560	27	81	−29	−52
	12	480	570	28	80	−69	−95
	22	490	560	27	81	−84	−106
	31	470	560	27	81	−74	−92
	57	500	570	25	78	−69	−88
	90	500	590	25	79	−63	−86
350	5	430	550	28	76	−28	−49
	15	480	560	25	80	−63	−83
	27	480	570	30	80	−64	−87
	50	520	590	27	79	−60	−82
	69	470	570	29	79	−59	−77
	110	480	580	27	78	−55	−77
410	4	430	520	29	78	−31	−57
	21	450	530	29	78	−61	−86
	35	480	560	26	76	−62	−93
	53	470	560	28	78	−59	−63
	69	480	560	27	78	−57	−63
	110	480	570	29	76	−52	−74

28 J transition temperatures have been plotted as well as the usual 100 J temperatures in Fig. 11.2. The change in transition temperature on stress relief (Fig. 11.3) was nearly always negative and was most marked for welds with no added titanium; the shift tended to be reduced as the titanium content was increased.

Although no metallography was carried out, several tentative explanations are proposed for some of the effects of oxygen. The strengthening effect may be a result of lower oxygen tying up less manganese and silicon as inclusions and thus making more of these elements available for strengthening. Reduction of sulphur content by the magnesium used to reduce oxygen may possibly have enhanced this effect.

The reduction in inclusion content as oxygen was lowered will have improved upper shelf toughness; this effect should have more effect on the 100 J than the 28 J transition. Because reducing oxygen

Table 11.2 (cont.) b) Stress-relieved at 580 °C.

O, ppm	Ti, ppm	Yield strength, N/mm²	Tensile strength, N/mm²	Elongation, %	Reduction of area, %	Charpy transition, °C	
						100 J	28 J
240	3	320	480	32	82	−81	−96
	16	350	500	31	82	−83	−104
	27	370	510	33	82	−93	−112
	35	400	530	34	80	−90	−109
	53	430	540	27	81	−82	−103
	97	440	550	28	80	−70	−99
280	5	340	490	32	81	−76	−100
	12	360	500	32	81	−80	−103
	22	400	520	32	80	−91	−118
	31	380	520	33	80	−92	−113
	57	430	540	30	79	−78	−105
	90	430	550	31	79	−72	−91
350	5	320	490	34	80	−76	−97
	15	330	500	35	80	−80	−103
	27	360	520	31	79	−80	−98
	50	370	530	30	79	−71	−99
	69	410	500	31	81	−66	−91
	110	440	530	32	79	−52	−83
410	4	330	470	32	79	−72	−93
	21	320	480	36	80	−80	−97
	35	350	500	33	79	−80	−100
	53	350	490	30	79	−74	−99
	69	360	500	32	80	−69	−87
	110	370	510	30	79	−55	−82

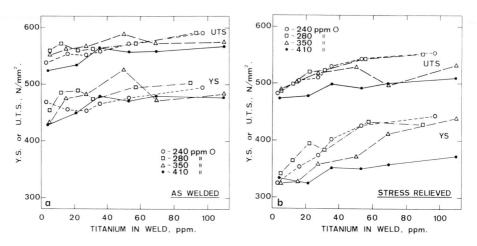

11.1 Influence of weld oxygen and titanium content on strength of 1.4%Mn weld metal: *a)* As-welded; *b)* Stress-relieved at 580 °C.

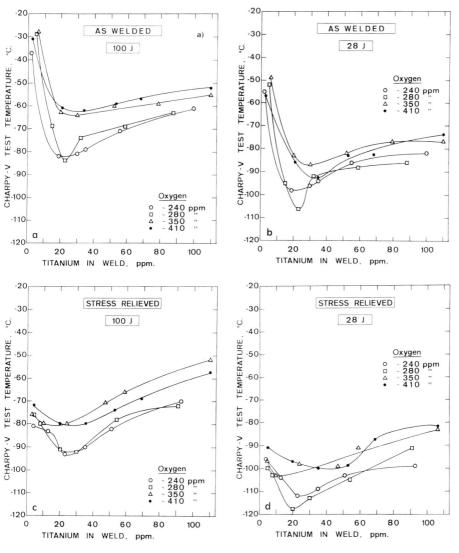

11.2 Influence of weld oxygen and titanium content on Charpy transition temperature of 1.4%Mn weld metal: *a*) 100 J as-welded; *b*) 28 J as-welded; *c*) 100 J stress-relieved at 580 °C; *d*) 28 J stress-relieved at 580 °C.

decreases the number of inclusions (rather than their size), less titanium will be required if it acts via inclusions. The effect cannot be a large one because as little as 30 ppm Ti (estimated by a less precise technique than used for the present work) has been reported[1] to give

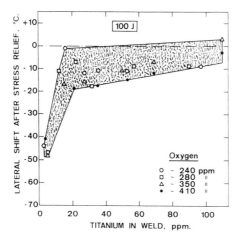

11.3 Change in 100 J Charpy transition temperature of 1.4%Mn weld metal with varying titanium and oxygen content on stress relief at 580 °C.

appreciable acicular ferrite in a manganese silicate submerged-arc weld metal with almost 1000 ppm oxygen.

Reducing the number of inclusions by reducing oxygen will eventually reduce the number to a level at which they are too few to achieve optimum nucleation of acicular ferrite and refinement of ferrite in the low temperature reheated regions; this level may have been reached at about 280 ppm O in the present tests.

Manganese and oxygen (Mn–O)

To study the joint influence of oxygen and manganese in the absence of titanium and other microalloying elements, welds were deposited at four levels of manganese (coded A_0 to D_0 because of the absence of titanium) and three oxygen levels.[2] Oxygen was deliberately varied below the normal level by additions of magnesium to the electrode coating. Analysed compositions of the welds are given in Table 11.3.

These deposits showed satisfactory low and/or constant levels of the invariant elements, except that both sulphur and nitrogen fell slightly with the magnesium added to the coating. There was no sensible decrease in oxygen content as the manganese content in each series was increased.

Table 11.3 Composition of weld metal having varying Mn and O content

Element, wt%		C	Mn	Si	S	P	N	O
Electrode	Oxygen level, %							
A_0	0.03	0.079	0.60	0.37	0.004	0.008	0.005	0.028
B_0	[4%Mg]	0.079	0.94	0.33	0.004	0.008	0.005	0.029
C_0		0.081	1.39	0.30	0.004	0.007	0.005	0.029
D_0		0.084	1.60	0.30	0.004	0.006	0.006	0.030
A_0	0.037	0.079	0.62	0.34	0.005	0.009	0.006	0.038
B_0	[2%Mg]	0.080	1.00	0.33	0.005	0.009	0.006	0.036
C_0		0.077	1.33	0.26	0.005	0.009	0.008	0.038
D_0		0.077	1.79	0.25	0.005	0.009	0.007	0.038
A_0	0.045	0.074	0.63	0.35	0.009	0.007	0.007	0.047
B_0	[0%Mg]	0.077	1.00	0.30	0.008	0.007	0.008	0.046
C_0		0.074	1.26	0.26	0.007	0.007	0.008	0.044
D_0		0.075	1.56	0.25	0.006	0.008	0.008	0.045

Notes: the values are the average for as-welded and stress-relieved deposits.
All welds contained <5 ppm Al, B, Nb, Ti, V, ~0.03%Cr, Ni and Cu, 0.005%Mo.

Metallography

In the top, as-deposited beads none of the welds contained more than about 25% acicular ferrite in its microstructure (Fig. 11.4), the primary constituent being ferrite with aligned second phase typified by the low oxygen D_0 deposit shown in Fig. 11.5. Increasing manganese reduced the proportion of primary ferrite, whilst reducing oxygen slightly increased the proportion of acicular ferrite.

Increasing manganese and, to a smaller degree, reducing oxygen, reduced the amount of primary ferrite in the coarse-grained regions and refined the intergranular microstructure, thus improving the definition of the prior austenite grain boundaries.

In the fine-grained regions, the two lowest manganese levels A_0 and B_0 gave equiaxed ferrite structures, whilst the other deposits gave a mixture of equiaxed ferrite and ferrite with second phase. Mean intercept ferrite grain sizes were between 4.6 and 6.0 μm, with no clear trends.

Etching in Le Pera's etch showed dark-etching carbides in the A_0 deposits on ferrite grain boundaries and at triple points. Increasing manganese (and decreasing oxygen at high manganese levels) increased the proportions of ferrite with second phase and of light-etching angular constituents. These were found to spheroidise on stress relief (Fig. 11.6).

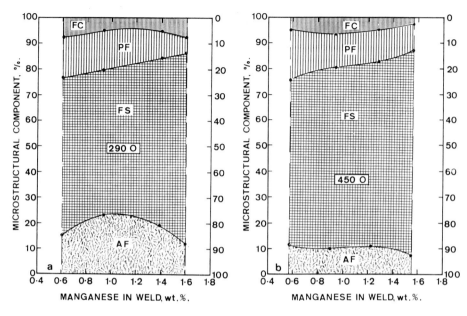

11.4 Influence of manganese on as-deposited microstructure of titanium-free deposit with: *a)* 290 ppm oxygen; *b)* 450 ppm oxygen.

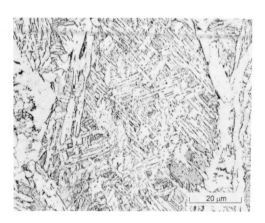

11.5 As-deposited microstructure of titanium-free deposit with 1.6%Mn and 0.029%O.

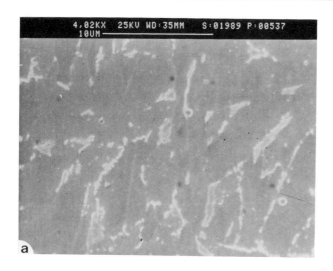

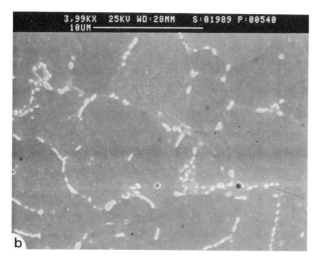

11.6 SEM micrographs of fine grained region of titanium-free deposits with 1.7%Mn and 0.045%O: *a*) As-welded; *b*) After stress relief at 580 °C.

Mechanical properties

In both as-welded and stress-relieved conditions manganese increased strength (Table 11.4), although the rates of increase were much less than with the standard (titanium-bearing) welds in Chapter 3, where (except for the lowest carbon levels) both yield and tensile strength increased by at least 100 N/mm^2 from the lowest to the highest

Table 11.4 Results of tensile tests on Ti-free welds of varying oxygen and manganese content

Electrode and state	O, ppm	Yield strength, N/mm²	Tensile strength, N/mm²	Elongation, %	Reduction of area, %	Charpy transition, °C 100 J	28 J
A_0, AW	290	400	490	29	80	−61	−77
B_0		400	490	28	79	−73	−88
C_0		440	540	25	80	−48	−64
D_0		450	530	n.d.	79	−40	−64
A_0, AW	370	390	480	28	79	−57	−75
B_0		420	500	30	79	−70	−95
C_0		430	520	26	79	−38	−58
D_0		470	560	26	78	−24	−53
A_0, AW	450	400	490	32	76	−46	−70
B_0		410	500	31	77	−56	−83
C_0		410	510	29	78	−37	−68
D_0		430	530	26	77	−11	−46
A_0, SR	290	300	440	33	81	−59	−74
B_0		380	450	34	82	−70	−85
C_0		330	480	33	81	−37	−99
D_0		360	490	33	82	−86	−106
A_0, SR	370	300	450	36	77	−54	−66
B_0		300	460	36	80	−68	−85
C_0		330	470	33	80	−76	−93
D_0		340	490	32	80	−74	−99
A_0, SR	450	330	450	30	79	−53	−79
B_0		370	480	30	79	−66	−90
C_0		310	460	36	80	−72	−87
D_0		310	480	34	80	−77	−110

Note: AW – as-welded; SR – stress relieved; n.d. not determined.

manganese levels (Tables 3.5 and 3.7). No consistent effect of oxygen was found. Stress relief gave a considerable, if inconsistent, decrease in strength values.

Regression analysis showed that, taking all three oxygen levels as belonging to the same set of results, the following linear relationships were obtained:

$$YS_{aw} = 361 + 52 \text{ Mn} \qquad [11.1]$$

$$TS_{aw} = 449 + 54 \text{ Mn} \qquad [11.2]$$

$$YS_{sr} = 314 + 15 \text{ Mn} \qquad [11.3]$$

$$TS_{sr} = 424 + 37 \text{ Mn} \qquad [11.4]$$

These equations all showed much lower manganese factors than the original Eq. [3.1]–[3.4], which were developed for the standard weld

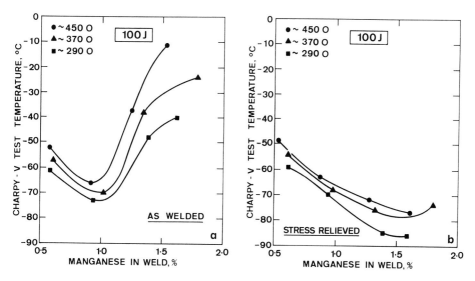

11.7 Influence of manganese on 100 J Charpy transition temperature of titanium-free deposits of varying oxygen content: *a)* As-welded; *b)* After stress relief at 580 °C.

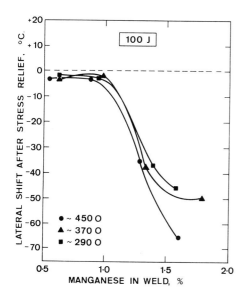

11.8 Influence of manganese on change in 100 J Charpy transition temperature on stress relief at 580 °C for titanium-free deposits of varying oxygen content.

metals containing ~50 ppm Ti, although the constant terms were similar. Correlation coefficients were all above 0.9, indicating a high degree of significance, except for Eq. [11.3], where the value of only 0.22 shows that manganese had no significant influence on the yield strength in the stress-relieved condition.

Charpy test results are summarised in Table 11.4 and in plots of 100 J transition temperature for different manganese content in Fig. 11.7. As-welded, the optimum manganese content for Charpy toughness is 1%, i.e. appreciably lower than the level of 1.4%Mn for weld metals containing some titanium described in Chapters 3 and 10. However, after PWHT, the optimum is closer to 1.5%, as in the weld metals with titanium. The changes in toughness on stress relief, shown in Fig. 11.8, are negligible at the two lowest manganese levels but rise to reductions of 35–70 °C with the C_0 and D_0 series.

Aluminium and oxygen (Al–O)

Excessive amounts of aluminium are known to interfere with formation of acicular ferrite in submerged-arc welds when titanium is present (Ref. 6, Ch. 10 and Ref. 8, Ch. 3). To explore this phenomenon in MMA welds, high purity 1.4%Mn welds were deposited from electrodes giving ~30 ppm Ti and up to 600 ppm Al in four sets of welds with normal and reduced oxygen levels, using magnesium in the coating to reduce the oxygen content down to approximately 310 ppm. Mechanical properties were determined in the as-welded condition only.

The analyses of the deposits in Table 11.5 show some slight but unintentional variations from the intended composition in all series except the standard oxygen level of 430 ppm, although most of the welds had compositions close to the intended. As with earlier welds with reduced oxygen, magnesium in the coating lowered sulphur and nitrogen as well as oxygen. Also given in Table 11.5 is the ratio of aluminium to oxygen, as this has been related to the ability to nucleate acicular ferrite and thus affect toughness (Ref. 2, Ch. 1).

The results of as-welded mechanical property tests are summarised in Table 11.6. Aluminium slightly increased strength and reduced ductility whereas oxygen had little effect. The major effect on yield strength appears to result from the first 100–200 ppm of aluminium, as shown in Fig. 11.9 (where the results are bounded by a scatter

Table 11.5 Composition (wt%) of 1.4%Mn weld metal with 35 ppm titanium and varying aluminium and oxygen content

Element	C	Mn	Si	S	P	Ti	N	O	Al	
O, ppm										Al/O ratio
320	0.074	1.51	0.35	0.004	0.009	0.0028	0.005	0.033	<0.0005	<0.02
	0.077	1.34	0.39	0.004	0.011	0.0030	0.005	0.031	0.0065	0.21
	0.083	1.35	0.47	0.004	0.009	0.0031	0.005	0.030	0.012	0.40
	0.094	1.25	0.46	0.004	0.010	0.0025	0.005	0.033	0.015	0.46
	0.080	1.26	0.46	0.004	0.011	0.0026	0.005	0.032	0.017	0.54
	0.103	1.34	0.61	0.004	0.009	0.0051	0.004	0.029	0.022	0.76
340	0.077	1.52	0.34	0.005	0.006	0.0033	0.006	0.032	0.0004	0.01
	0.082	1.44	0.36	0.004	0.006	0.0039	0.006	0.033	0.0049	0.15
	0.084	1.39	0.42	0.004	0.006	0.0044	0.005	0.034	0.0095	0.28
	0.082	1.35	0.47	0.005	0.006	0.0044	0.005	0.034	0.015	0.43
	0.086	1.30	0.50	0.004	0.007	0.0032	0.006	0.033	0.024	0.72
	0.087	1.28	0.58	0.004	0.006	0.0032	0.005	0.035	0.035	1.01
400	0.074	1.48	0.35	0.006	0.007	0.0034	0.008	0.040	<0.0005	<0.02
	0.077	1.50	0.37	0.005	0.007	0.0053	0.007	0.041	0.0069	0.17
	0.079	1.44	0.44	0.005	0.007	0.0042	0.005	0.041	0.016	0.39
	0.078	1.39	0.53	0.005	0.007	0.0045	0.005	0.041	0.028	0.68
	0.077	1.40	0.61	0.004	0.008	0.0052	0.005	0.039	0.043	1.10
	0.075	1.35	0.67	0.004	0.007	0.0049	0.005	0.039	0.048	1.39
430	0.069	1.36	0.30	0.007	0.009	0.0037	0.007	0.043	<0.0005	<0.02
	0.080	1.41	0.33	0.006	0.009	0.0043	0.006	0.043	0.0078	0.18
	0.076	1.36	0.37	0.005	0.009	0.0036	0.005	0.042	0.019	0.45
	0.078	1.31	0.44	0.005	0.008	0.0036	0.005	0.043	0.034	0.79
	0.076	1.30	0.51	0.005	0.008	0.0043	0.005	0.042	0.049	1.16
	0.079	1.32	0.57	0.004	0.008	0.0038	0.005	0.042	0.061	1.45

Notes: all welds contained <5 ppm B, Nb, V, ~0.03%Cr, Ni and Cu, 0.005%Mo.
The Al/O ratio for Al_2O_3 is 1.124.

band). However, as this was accompanied (Table 11.5) by an increase in silicon, the latter may have been the actual strengthening agent.

A plot of 100 J Charpy impact temperatures (Fig. 11.10) shows that the first addition of 50–80 ppm of aluminium was harmful. Subsequent increases to 150–300 were beneficial to a greater or lesser degree but further increases were again harmful. With aluminium contents below about 250 ppm, low oxygen contents were beneficial but above this level of aluminium, low oxygen appeared to be detrimental. The worst toughness of 100 J at −38 °C was found with weld metal containing 610 ppm Al and 420 ppm O; it was as poor as some of the titanium-free welds in Tables 11.2 and 11.4. The results do not appear to support the hypotheses that the Al:O ratio is important, or that the ability to achieve good toughness depends on the value of

Table 11.6 Mechanical properties, as-welded, for 1.4%Mn welds with 35 ppm Ti and varying aluminium and oxygen content

O, ppm	Al, ppm	Yield strength, N/mm²	Tensile strength, N/mm²	Elongation, %	Reduction of area, %	Charpy transition, °C 100 J	28 J
320	<5	450	530	29	82	−80	−92
	65	480	550	30	82	−72	−87
	120	490	560	27	82	−80	−94
	150	490	560	26	82	−83	−96
	170	480	550	27	81	−76	−95
	220	480	550	29	79	−65	−84
340	4	480	550	28	81	−77	−94
	49	480	550	30	79	−65	−85
	95	500	550	27	78	−73	−86
	150	490	550	29	81	−78	−96
	240	500	560	26	81	−64	−83
	350	490	570	27	79	−50	−78
400	<5	460	530	30	79	−69	−93
	69	480	550	28	79	−59	−80
	160	500	560	32	80	−61	−83
	280	480	560	29	79	−63	−84
	430	500	570	28	78	−46	−84
	480	510	580	27	76	−43	−77
430	<5	470	530	33	81	−72	−99
	78	480	540	31	81	−55	−77
	190	480	550	31	80	−58	−86
	340	490	560	28	79	−63	−88
	490	480	560	25	79	−55	−90
	610	490	570	29	77	−38	−74

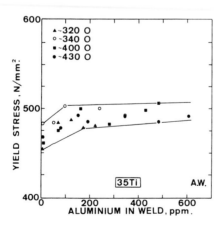

11.9 Influence of weld oxygen and aluminium content on as-welded yield strength of 1.4%Mn weld metal containing ~35 ppm Ti.

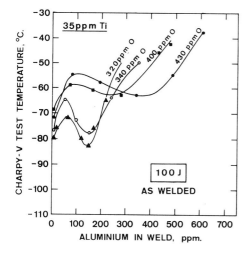

11.10 Influence of weld oxygen and aluminium content on 100 J Charpy transition temperature of 1.4%Mn weld metal with ~35 ppm Ti.

oxygen in excess of what is needed to combine with any aluminium present.

Discussion

It is widely known that oxygen, as a promoter of non-metallic inclusions, reduces resistance to ductile fracture, i.e. it reduces upper shelf Charpy toughness. Its role in controlling cleavage fracture and the transition between the two modes of fracture is less certain and inevitably more complex. In the absence of complicating factors, reducing the oxygen content slightly improves the 100 J transition temperature.

The main effects of oxygen are indirect, although oxygen is certainly needed to achieve a suitable inclusion population, i.e. probably depending mainly on inclusions having titanium oxide at their surfaces. It is possible, from the Ti–O series, that reducing oxygen much below 280 ppm is likely to reduce the number of sites for nucleation of acicular ferrite below the optimum. Aluminium, when present, can interfere with this nucleation – possibly because it prevents formation of titanium oxide – and it is likely that the lower the oxygen content the less aluminium would be required to interfere.

References

1 Bailey N: 'Ferritic steel weld metal microstructures and properties'. Proc conf on 'Perspectives in metallurgical developments'. Sheffield University, 1984, Metals Soc, London, 1984.
2 Abson DJ and Evans GM: 'A study of the Mn–O system in low-hydrogen MMA all-weld-metal deposits'. Proc 2nd int conf, Gatlinburg, USA, ASM Intl, 1990, 821–25; IIW Doc II–A–770–89.

12

Combinations with nitrogen

Effects of nitrogen, as an alloying element and as a promoter of strain ageing embrittlement, were discussed for weld metals of relatively low purity in Chapter 7. The present chapter considers these two aspects in relation to high purity weld metals, both without titanium and with different levels of the element. It also deals with effects of aluminium which, in bulk steels, is used to control strain ageing. Attempts to develop a weld metal with a low propensity to strain ageing are presented and, finally, possibilities for reducing nitrogen content during welding are discussed.

Titanium and nitrogen (Ti–N)

The joint influences of titanium and nitrogen were examined for welds with 1.5%Mn,[1,2] whose compositions are given in Table 12.1. As with other welds with titanium in Chapters 10 and 11, increasing titanium additions led to increasing silicon and decreasing oxygen content; other elements remained sensibly constant throughout.

Metallography

As with the less pure welds in Chapter 7, nitrogen had no effect on the macrostructure of the welds examined. Titanium had an overwhelming influence on microstructure, and its effects on welds of standard nitrogen content are included in Chapter 10, in the section 'Titanium additions to high purity weld metal'.

Increasing the nitrogen content of titanium-free welds (series O) gave coarser microstructures with a significant increase in the amount of ferrite with second phase, as shown in Fig. 12.1(a,b). In comparison with the acicular ferritic welds of low titanium content (Fig. 12.1(c,d)), welds of high titanium content (with both high and low nitrogen content) were essentially similar (Fig. 12.1(e,f)), except that

Table 12.1 Composition of 1.4%Mn weld metal having varying nitrogen and titanium content

Element, wt%		C	Mn	Si	S	P	Ti	N	O
N, ppm	Code								
80	O	0.074	1.40	0.25	0.008	0.007	0.0001	0.0079	0.048
	W	0.077	1.46	0.27	0.008	0.007	0.0028	0.0081	0.046
		0.073	1.41	0.26	0.007	0.011	0.009	0.0083	0.039
		0.074	1.45	0.30	0.007	0.009	0.012	0.0084	0.034
		0.072	1.50	0.38	0.006	0.010	0.026	0.0077	0.029
	X	0.069	1.47	0.45	0.005	0.006	0.041	0.0077	0.028
		0.070	1.50	0.45	0.005	0.006	0.055	0.0075	0.029
150	O1	0.074	1.58	0.28	0.008	0.008	<0.0005	0.0145	0.040
	W1	0.068	1.40	0.28	0.010	0.008	0.0031	0.0148	0.041
		0.069	1.50	0.31	0.008	0.008	0.0051	0.0160	0.040
		0.070	1.50	0.29	0.008	0.009	0.012	0.0164	0.034
		0.068	1.51	0.39	0.007	0.009	0.030	0.0166	0.028
	X1	0.066	1.48	0.47	0.007	0.011	0.041	0.0164	0.028
		0.070	1.50	0.45	0.004	0.006	0.059	0.0155	0.030
240	O2	0.073	1.66	0.27	0.009	0.008	<0.0005	0.0235	0.040
	W2	0.069	1.45	0.26	0.010	0.009	0.0029	0.0226	0.039
		0.073	1.53	0.29	0.009	0.007	0.0046	0.0243	0.042
		0.070	1.45	0.28	0.009	0.009	0.012	0.0239	0.032
		0.067	1.48	0.40	0.006	0.008	0.032	0.0253	0.029
	X2	0.068	1.46	0.47	0.007	0.006	0.045	0.0249	0.030
		0.066	1.47	0.43	0.005	0.006	0.069	0.0240	0.032

Note: all welds contained <5 ppm B, with ~0.03% each of Cr, Cu and Ni, 0.005%Mo.

they contained more ferrite with aligned second phase originating in the primary ferrite at the prior austenite grain boundaries.

Mechanical properties

Tensile properties (Table 12.2) in the as-welded condition showed that strength increased somewhat irregularly with increasing titanium content (Fig. 12.2) and that, although the first increase in nitrogen content increased strength levels noticeably, the second increase up to 240 ppm N had little further effect.

Charpy tests were carried out on all deposits stress-relieved as well as as-welded and the results, expressed as 100 J transition temperatures, are included in Table 12.2 and plotted in Fig. 12.3. Nitrogen impaired toughness in all series of welds; as-welded the increase in 100 J transition temperature was just over 40 °C. Stress relief lowered transition temperatures of all deposits but, in contrast to the 'stand-

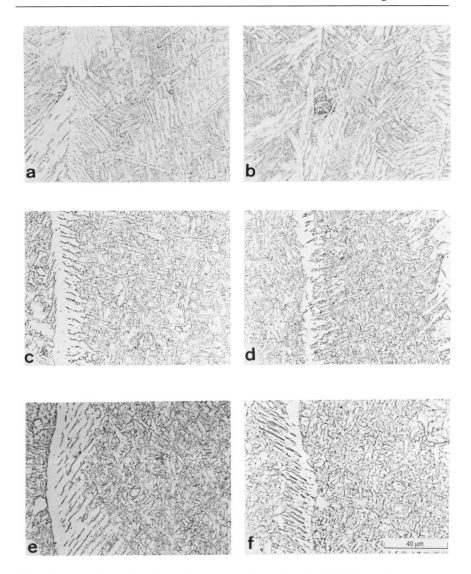

12.1 Microstructure of as-deposited regions of welds with 1.4%Mn and:
a) <5 ppm Ti, 80 ppm N; b) <5 ppm Ti, 240 ppm N; c) 35 ppm Ti, 80 ppm N;
d) 35 ppm Ti, 240 ppm N; e) 400 ppm Ti, 80 ppm N; f) 400 ppm Ti,
240 ppm N.

ard' welds described in Chapter 7, nitrogen was almost as detrimental in the stress-relieved condition as it was as-welded.

Fracture toughness CTOD tests on specimens from selected welds (the results of which are summarised in Fig. 12.4) essentially

Table 12.2 Mechanical properties, as-welded except for charpy results, of 1.4%Mn welds of varying nitrogen and titanium content

N, ppm	Code	Yield strength, N/mm²	Tensile strength, N/mm²	Elongation, %	Reduction of area, %	Charpy transition, °C 100 J AW	Charpy transition, °C 100 J SR
80	O	440	530	28	78	−14	−74
	W	470	540	25	77	−68	−70
		480	540	24	79	−50	nd
		480	540	28	80	−43	−56
		510	580	26	79	−56	−67
	X	500	580	26	80	−61	−69
		530	600	25	82	−60	−63
150	O1	470	570	24	77	5	−45
	W1	480	540	28	79	−41	−62
		520	580	26	79	−46	−62
		510	570	25	76	−29	−54
		550	610	24	78	−35	−58
	X1	580	640	23	75	−44	−58
		540	600	26	76	−41	−55
240	O2	500	610	24	76	20	−29
	W2	490	580	26	78	−24	−50
		540	590	25	74	−28	−50
		520	600	29	77	−23	−37
		550	620	24	76	−24	−46
	X2	580	630	25	78	−30	−43
		550	620	26	78	−28	−48

Note: AW – as-welded; SR – stress relieved; nd – not determined.

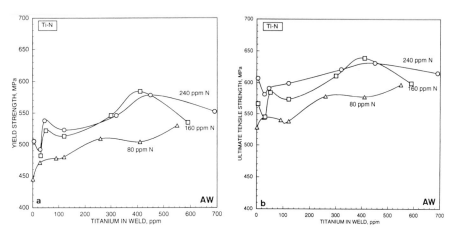

12.2 Influence of weld nitrogen content of 1.4%Mn welds of varying titanium level on as-welded: *a*) Yield strength; *b*) Tensile strength.

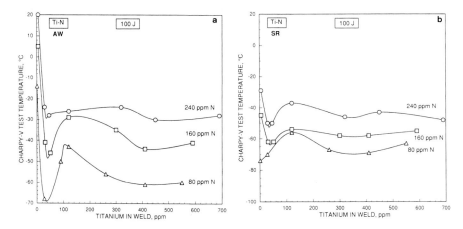

12.3 Influence of weld nitrogen content of 1.4%Mn welds of varying titanium level on 100 J Charpy transition temperature: *a*) As-welded; *b*) Stress relieved at 580 °C.

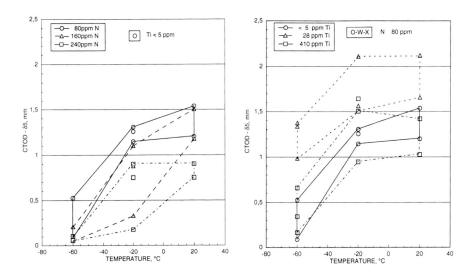

12.4 Results of CTOD tests on 1.4%Mn welds of varying nitrogen and titanium content in as-welded condition: *a*) Varying N at fixed Ti level; *b*) Varying Ti at fixed N. Specimen cross section 17 × 34 mm, through-thickness notch at mid-weld position.

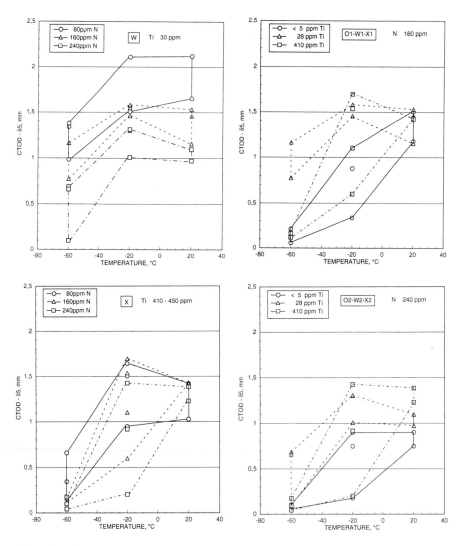

12.4 *(Cont.)*

confirmed the findings of the Charpy tests, except that deposits with 400 ppm Ti were always inferior to those with 30 ppm Ti, regardless of nitrogen content.

Aluminium and nitrogen (Al–N)

Aluminium is the element used in killed steels to complete deoxidation; the excess then combines with nitrogen and thus controls strain

Table 12.3 Composition of 1.4%Mn, 35 ppm Ti weld metal having varying nitrogen and aluminium content

Element, wt%	C	Mn	Si	S	P	Ti	Al	N	O
N, ppm									
60	0.069	1.36	0.30	0.007	0.009	0.0037	0.0005	0.0068	0.043
	0.073	1.39	0.33	0.007	0.008	0.0039	0.0020	0.0066	0.043
	0.078	1.39	0.32	0.007	0.009	0.0042	0.0044	0.0066	0.044
	0.080	1.41	0.33	0.006	0.009	0.0043	0.0078	0.0055	0.043
	0.080	1.42	0.37	0.006	0.009	0.0043	0.012	0.0054	0.044
	0.076	1.36	0.37	0.005	0.009	0.0036	0.019	0.0053	0.042
	0.078	1.31	0.44	0.005	0.008	0.0036	0.034	0.0050	0.043
	0.076	1.30	0.51	0.005	0.008	0.0043	0.049	0.0048	0.042
	0.079	1.32	0.57	0.004	0.008	0.0038	0.061	0.0052	0.042
140	0.069	1.35	0.32	0.010	0.006	0.0030	0.0005	0.0149	0.044
	0.070	1.40	0.32	0.009	0.007	0.0037	0.0069	0.0159	0.045
	0.069	1.37	0.38	0.008	0.007	0.0033	0.014	0.0141	0.045
	0.073	1.29	0.47	0.006	0.005	0.0031	0.028	0.0133	0.048
	0.072	1.32	0.50	0.006	0.006	0.0032	0.044	0.0132	0.041
	0.070	1.30	0.54	0.005	0.006	0.0029	0.058	0.0119	0.043
200	0.070	1.39	0.28	0.009	0.010	0.0031	0.0002	0.0242	0.045
	0.070	1.41	0.35	0.007	0.007	0.0038	0.0084	0.0214	0.044
	0.070	1.37	0.39	0.007	0.009	0.0032	0.021	0.0202	0.048
	0.070	1.36	0.42	0.006	0.007	0.0036	0.038	0.0200	0.046
	0.076	1.42	0.49	0.005	0.010	0.0055	0.047	0.0191	0.036
	0.070	1.40	0.54	0.005	0.009	0.0044	0.067	0.0195	0.036

Note: all welds contained ~0.03% each of Cr, Cu and Ni, 0.006%Mo, ≤5 ppm each of Nb and V.

ageing. In weld metals, less time is available to complete the deoxidation process, so that they contain more oxygen and much, if not all, of the aluminium measured by chemical analysis is in the form of the oxide Al_2O_3, either on its own or as a mixed oxide. Only oxygen in excess of that needed to form Al_2O_3 is available to combine with nitrogen. Nevertheless, it is important to know the combined effects of aluminium and nitrogen to understand the possibilities for control of strain ageing.

Electrodes were made to give 1.4%Mn welds with about 35 ppm Ti, with controlled aluminium content up to 670 ppm and low levels of other microalloying elements, all at three nitrogen levels. The analyses of the resultant welds, given in Table 12.3, show that silicon increased, sulphur decreased, and nitrogen slightly decreased as aluminium was increased; otherwise compositions were well balanced.

Mechanical properties

The results of as-welded tensile tests, together with 100 J Charpy transition temperature (the latter both as-welded and after stress relief at 580 °C), are presented in Table 12.4. The effect of aluminium on

Table 12.4 Mechanical properties, as-welded except for Charpy, of 1.4%Mn welds with 36 ppm Ti and varying nitrogen and aluminium content

N, ppm	Al, ppm	Yield strength, N/mm^2	Tensile strength, N/mm^2	Elongation, %	Reduction of area, %	Charpy transition, °C	
						100 J AW	100 J SR
60	5	470	530	33	81	−72	−78
	20	480	540	33	81	−70	−78
	44	480	540	30	79	−60	−75
	78	480	540	31	81	−55	−68
	120	480	540	29	79	−58	−69
	190	480	550	31	80	−58	−74
	340	490	560	28	79	−63	−74
	490	480	560	25	79	−55	−69
	610	490	570	29	77	−38	−52
140	5	490	570	30	77	−45	−60
	69	500	570	26	79	−39	−50
	140	520	590	26	79	−42	−58
	280	510	580	24	77	−38	−59
	470	510	580	25	76	−30	−56
	580	510	580	26	76	−15	−55
200	2	520	600	25	75	−27	−46
	84	550	620	27	76	−27	−44
	210	560	630	27	78	−30	−51
	380	550	620	30	75	−21	−47
	470	540	620	28	72	0	−37
	760	540	620	28	74	2	−34

Note: AW – as-welded; SR – stress relieved.

these properties at the low nitrogen level is considered in more detail in Chapter 13. However, Fig. 12.5 shows a slight but irregular strengthening effect of aluminium. Increasing nitrogen increased both yield and tensile strength, the increase being greater between 140 and 200 ppm N than between 60 and 140 ppm. At the two higher nitrogen levels, a distinct peak in strength occurred between 100 and 200 ppm Al.

Charpy temperatures, plotted in Fig. 12.6, increased with increasing nitrogen content both as-welded and stress-relieved. In both conditions, the increase was less between 140 and 200 ppm N than between 60 and 140 ppm, i.e. roughly in line with the differences between nitrogen levels. Stress relief was always beneficial, the benefit increasing as both nitrogen and aluminium were increased. At the highest nitrogen level, increasing aluminium from 470 to 670 ppm did not further increase the transition temperature, although similar

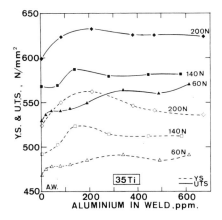

12.5 Influence of aluminium on as-welded strength of 1.4%Mn welds with ~35 ppm Ti at three nitrogen levels.

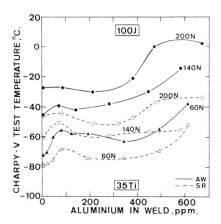

12.6 Influence of aluminium content of 1.4%Mn welds with ~35 ppm Ti at three nitrogen levels on 100 J Charpy transition temperature, as-welded and stress relieved at 580 °C.

increases sharply raised transition temperatures for deposits with 140 ppm N (as-welded only) and with 60 ppm N in both conditions.

Although strain ageing embrittlement tests were not carried out on these deposits, the prospect of using aluminium to minimise strain ageing problems can be assessed from the compositions available. Such a weld metal would require sufficient aluminium in solid solution to combine with the nitrogen present to form aluminium nitride,

AlN. Of the welds in Table 12.3, only three fulfilled this condition; they were the welds of the highest aluminium content at each nitrogen level.

Their as-welded 100 J Charpy transition temperatures were −38, −15 and +2 °C, respectively. Such values are not good enough for the type of service required of high quality basic C–Mn electrodes and it must be concluded that the only hope for such an approach might be to select weld metals of such low nitrogen and oxygen content that less aluminium would be needed to be able to form AlN. However, even with the reduced oxygen (290 ppm) series in Table 11.5, the weld having the highest aluminium content of 220 ppm had insufficient aluminium for any to be uncombined with oxygen, leaving none to form AlN; this weld gave a 100 J transition temperature of −65 °C, higher than the best in this series and a value which would probably worsen were sufficient aluminium added to enable AlN to be formed.

Table 12.5 Composition of 1.4%Mn weld metal having varying boron and nitrogen content

Element, wt%		C	Mn	Si	S	P	Ti	B	N	O
N, ppm	Ti, ppm									
160	<5	0.074	1.58	0.28	0.008	0.008	0.0007	0.0005	0.015	0.040
		0.074	1.60	0.31	0.007	0.008	0.0005	0.0015	0.013	0.041
		0.077	1.55	0.32	0.007	0.008	0.0005	0.0038	0.014	0.040
		0.074	1.59	0.35	0.007	0.008	0.0005	0.0072	0.016	0.037
		0.075	1.58	0.33	0.007	0.008	0.0005	0.0109	0.015	0.041
		0.082	1.60	0.36	0.007	0.008	0.0005	0.0141	0.014	0.036
		0.070	1.55	0.39	0.007	0.007	0.0005	0.0160	0.014	0.040
160	40	0.068	1.40	0.28	0.010	0.009	0.0031	0.0002	0.015	0.041
		0.056	1.37	0.26	0.006	0.009	0.0036	0.0038	0.013	0.036
		0.050	1.40	0.28	0.007	0.010	0.0040	0.0080	0.013	0.039
		0.048	1.36	0.31	0.007	0.010	0.0045	0.0119	0.013	0.030
		0.048	1.38	0.32	0.006	0.009	0.0045	0.0161	0.013	0.033
		0.045	1.40	0.35	0.006	0.009	0.0047	0.0190	0.013	0.033
240	<5	0.073	1.66	0.27	0.009	0.008	0.0005	0.0005	0.024	0.040
		0.078	1.60	0.30	0.007	0.007	0.0005	0.0030	0.021	0.039
		0.083	1.59	0.31	0.007	0.008	0.0005	0.0042	0.023	0.038
		0.079	1.58	0.34	0.007	0.007	0.0005	0.0080	0.022	0.036
		0.071	1.60	0.36	0.007	0.008	0.0005	0.0111	0.023	0.038
		0.078	1.58	0.38	0.007	0.007	0.0005	0.0123	0.023	0.034
		0.064	1.43	0.36	0.008	0.008	0.0005	0.0160	0.023	0.043
240	40	0.069	1.45	0.26	0.010	0.013	0.0029	0.0002	0.023	0.039
		0.048	1.42	0.24	0.006	0.008	0.0035	0.0034	0.023	0.031
		0.048	1.45	0.26	0.006	0.009	0.0038	0.0066	0.024	0.031
		0.047	1.44	0.29	0.006	0.008	0.0041	0.0105	0.024	0.029
		0.045	1.47	0.34	0.006	0.009	0.0045	0.0167	0.022	0.031
		0.046	1.46	0.34	0.006	0.009	0.0046	0.0197	0.022	0.035

Notes: for compositions of welds with 80 ppm N see Table 13.7.
All welds contained ~0.03%Cr, Ni and Cu, 0.005%Mo, <5 ppm Al, Nb and V.

Boron and nitrogen (B–N)

Boron and nitrogen can combine strongly to form boron nitride and it was important to know whether this could be used to any advantage either in the absence or the presence of a small (~40 ppm) titanium addition.

Electrodes were therefore made and tested[3] to produce 1.4%Mn welds with up to 200 ppm B at three levels of nitrogen and low content of other microalloying elements, except for the 40 ppm Ti added to two of the series. The compositions of the welds with the lowest level of nitrogen are given later in Table 13.7, the others in

Table 12.6 Mechanical properties of 1.4%Mn welds with varying boron and nitrogen content, with and without 40 ppm Ti

N/Ti	B, ppm	Yield strength, N/mm^2	Tensile strength, N/mm^2	Elongation, %	Reduction of area, %	100 J Charpy transition, °C	
						AW	SR
160	5	470	570	24	77	5	−45
<5	15	480	560	26	73	6	−40
	38	490	570	25	72	6	−36
	72	470	570	25	72	−2	−38
	109	460	540	25	76	−16	−48
	141	470	550	25	73	−29	−38
	160	470	560	21	67	−52	−46
160	2	480	540	28	79	−41	−62
40	38	460	530	26	74	−39	−59
	80	470	530	27	78	−23	−47
	119	430	500	27	78	−32	−53
	161	430	490	25	74	−32	−55
	190	450	510	25	77	−38	−43
240	5	500	610	24	76	20	−29
<5	30	510	600	23	72	18	−21
	42	500	590	23	73	15	−29
	80	490	580	23	71	13	−18
	111	480	560	24	72	8	−26
	123	470	560	25	73	1	−47
	160	440	530	22	71	−13	−40
240	2	490	580	26	78	−24	−50
40	34	500	570	25	71	−28	−39
	66	480	550	28	75	−13	−25
	105	460	530	28	76	−16	−28
	167	460	520	27	73	−23	−36
	197	430	500	27	75	−26	−39

Notes: for properties of welds with 80 ppm N see Table 13.8.
AW – as-welded; SR – stress-relieved.

Table 12.5. The compositions were well balanced except for a slight increase in silicon as the boron content was increased.

Mechanical properties

Mechanical properties, given in Table 12.6 and 13.8, are for as-welded tensile tests and for Charpy tests as-welded and after stress relief at 580 °C. In the absence of titanium, nitrogen had a strengthening influence with boron levels up to about 120 ppm. This influence was lost at higher contents of boron and, at 160 ppm B, the weld with 240 ppm N was the weakest (Fig. 12.7). With 40 ppm Ti present, nitrogen was no longer a strengthening element when boron was present, although the intermediate nitrogen level generally gave the lowest yield strength value.

Charpy 100 J temperatures (Table 12.6 and Fig. 12.8), show nitrogen to be an embrittling element in the presence of boron. One exception in the as-welded condition is the titanium-free weld with 160 ppm N at the highest boron level of 160 ppm. Neither this weld, nor the other welds with increased nitrogen, show any signs of the slight fall in toughness at the highest boron content of 200 ppm which was apparent in the deposits with a normal nitrogen level. In fact, the welds with titanium and enhanced nitrogen showed a continuing improvement in toughness as boron was increased from ~160 ppm to 200 ppm.

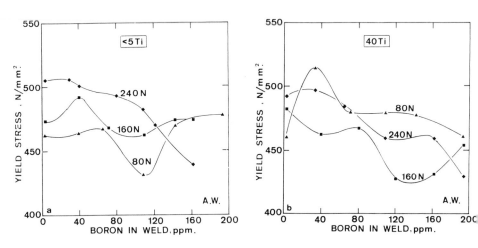

12.7 Influence of weld boron content on yield strength of as-welded 1.4%Mn welds at varying nitrogen levels with: a) <5 ppm Ti; b) 40 ppm Ti.

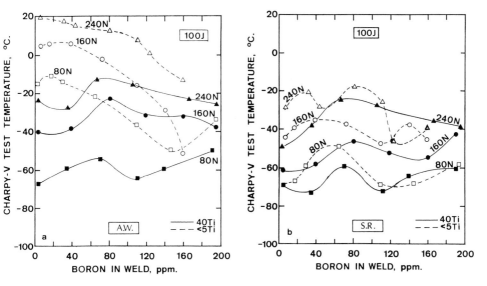

12.8 Influence of weld boron content on 100 J Charpy transition temperature of 1.4%Mn deposits with and without 40 ppm Ti at varying nitrogen levels: *a*) As-welded; *b*) Stress relieved at 580 °C.

In the absence of titanium, boron improved toughness up to 160 ppm, regardless of nitrogen level. After stress relief (Fig. 12.8(*b*)), changes were less dramatic although increasing nitrogen was nearly always accompanied by an increase in transition temperature, whilst boron, overall, had a negligible effect.

Static strain ageing

Selected welds (those coded in the O, W and X series in Tables 12.1 and 12.2) were examined in the statically strain aged condition (Ref. 5, Ch. 10 and Ref. 1 this chapter). The procedure, shown schematically in Fig. 7.1, was used to provide statically strain aged specimens for testing, with 4% compression (instead of 10%) at ambient temperature and aging for 0.5 hour at 250 °C. Specimens were then tested after strain ageing for comparison with previous results in the as-welded condition (Table 12.2).

Mechanical properties

The results of low load hardness testing are summarised in Fig. 12.9. The uniformity of hardness before and after ageing can be seen in

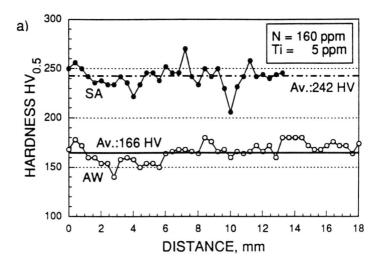

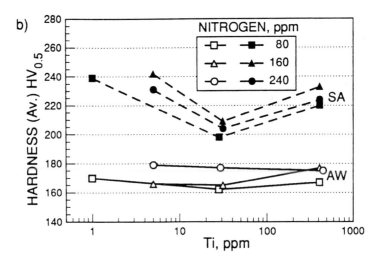

12.9 Low load hardness tests on 1.4%Mn welds with varying levels of Ti and N to show effect of strain ageing (4% compression, aged 0.5 hr/250 °C): a) Hardness distribution before and after strain ageing a low-Ti, medium-N weld metal; b) Influence of Ti and N on mean hardness before and after strain ageing (AW – as-welded; SA – strain aged).

Fig. 12.9(a) for a weld with <5 ppm Ti and 160 ppm N. As-welded (Fig. 12.9(b)) hardness values were between 160 and 180 HV 0.5, with little influence of either titanium or nitrogen. After strain ageing, the deposits with about 30 ppm Ti were distinctly softer, suggesting

Table 12.7 Influence of strain ageing on transition temperature of C–1.4%Mn weld metals with varying nitrogen and titanium content

Electrode and compositional type, ppm	100 J Charpy transition temperature, °C		Increase of transition temp, °C
	As-welded	Strain aged*	
O, 1 Ti, 79 N	−14	9	23
O1, <5 Ti, 145 N	5	38	33
O2, <5 Ti, 235 N	20	29	9
W, 28 Ti, 81 N	−68	−38	30
W1, 31 Ti, 148 N	−41	−20	21
W2, 29 Ti, 226 N	−24	−6	18
X, 410 Ti, 77 N	−61	−47	14
X1, 410 Ti, 164 N	−44	−24	20
X2, 450 Ti, 249 N	−30	−12	18

Note: * 4% compression at ambient temperature + 0.5 hr/250 °C.

that they might be somewhat less susceptible to strain ageing than either the Ti-free or the high titanium welds. In all conditions, welds with normal nitrogen levels of 80 ppm were the softest, although after strain ageing those with 240 ppm N were not as hard as those with 160 ppm.

No tensile testing was carried out after strain ageing (the as-welded properties can be found in Table 12.2, which also includes the results of as-welded Charpy tests). From the summary of Charpy temperatures in Table 12.7 and Fig. 12.10, it can be seen that strain ageing led to increases in transition temperature of between about 10 and 30 °C. In contrast to the indications from the hardness tests (Fig. 12.9), the titanium-free welds appeared to be of similar susceptibility to the 30 ppm Ti deposits, both being somewhat inferior to the welds of highest titanium content.

Crack tip opening displacement (CTOD) tests were carried out at −20 °C on selected welds in the strain-aged condition at three nitrogen levels (Fig. 12.11). This temperature was selected as being within the transition range in the as-welded condition (Fig. 12.4), when most tests gave maximum load CTOD values, except at the highest nitrogen level. After strain ageing, CTOD values fell sharply and only the 30 ppm Ti welds at the two lower nitrogen levels resulted in three maximum load values out of three tests. Although in Chapter 10 it was demonstrated that titanium levels around 400 ppm are capable of giving as-welded Charpy toughness as good as those with 30 ppm Ti, the welds of the lower titanium content were tougher in both

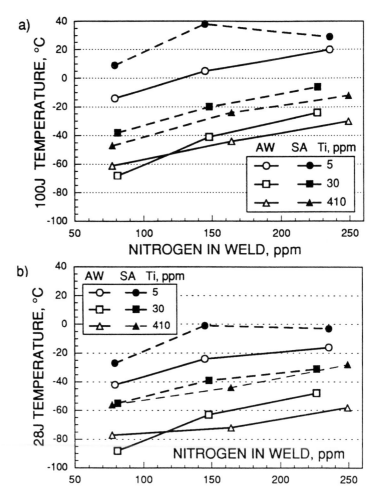

12.10 Effect of nitrogen and strain ageing (4% compression) on *a)* 100 J; *b)* 28 J Charpy transition temperature, for 1.4%Mn weld metals of varying titanium content.

as-welded and strain-aged conditions when assessed with fracture mechanics tests.

Dynamic strain ageing with selected additions

In the project to examine the feasibility of developing electrodes with improved resistance to strain ageing in service at moderately elevated temperatures,[4] electrodes were selected from previous projects alloyed

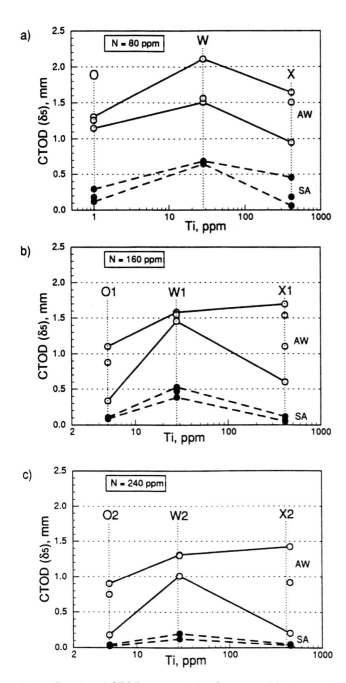

12.11 Results of CTOD tests at −20 °C on 1.4%Mn welds of varying titanium and nitrogen content as-welded and strain aged (4% compression), welds with: a) 80 ppm N; b) 160 ppm N; c) 240 ppm N. Specimen cross section 17 × 34 mm, through-thickness notch at mid-weld position.

Table 12.8 Weld composition for dynamic strain ageing tests

Element, wt%	C	Mn	Si	Al	Ti	B	N_t	O
Series								
Carbon	0.045	1.38	0.35	0.0013	0.0057	nd	0.0090	0.045
	0.123	1.43	0.41	0.0006	0.0071	nd	0.0091	0.035
Manganese	0.074	0.52	0.36	0.0005	0.0072	0.0006	0.0075	0.044
	0.085	0.97	0.38	0.0005	0.0061	0.0008	0.0082	0.040
[Baseline]	0.074	1.47	0.37	0.0005	0.0055	0.0007	0.0098	0.038
	0.077	1.85	0.34	0.0005	0.0047	0.0007	0.0120	0.040
Silicon	0.068	1.49	0.17	0.0005	0.0047	nd	0.0105	0.045
	0.077	1.58	0.69	0.0005	0.0065	nd	0.0090	0.035
	0.072	1.52	1.10	0.0003	0.0094	nd	0.0080	0.033
Aluminium	0.079	1.58	0.36	0.0016	0.0066	nd	0.0090	0.042
	0.085	1.58	0.36	0.0053	0.0076	nd	0.0080	0.042
	0.093	1.53	0.38	0.0085	0.0081	nd	0.0067	0.043
	0.085	1.48	0.41	0.0130	0.0063	nd	0.0065	0.045
Titanium	0.081	1.56	0.31	0.0005	0.0007	nd	0.0081	0.040
	0.077	1.55	0.32	0.0004	0.0030	nd	0.0105	0.040
	0.084	1.56	0.40	0.0006	0.012	nd	0.0095	0.035
	0.076	1.52	0.41	0.0007	0.016	nd	0.0092	0.034
Boron–1	0.073	1.44	0.41	0.0009	0.0066	0.0035	0.0089	0.039
	0.074	1.43	0.41	0.0005	0.0065	0.0054	0.0098	0.040
	0.070	1.41	0.43	0.0006	0.0065	0.0069	0.0096	0.041
Boron–2	0.070	1.47	0.36	<0.0005	0.0037	0.0001	0.0085	0.044
	0.067	1.47	0.38	<0.0005	0.0038	0.0028	0.0082	0.041
	0.062	1.41	0.35	<0.0005	0.0036	0.0046	0.0079	0.043
	0.062	1.39	0.37	<0.0005	0.0037	0.0063	0.0082	0.044
Nitrogen	0.075	1.45	0.36	0.0005	0.0063	nd	0.0170	0.040
	0.080	1.46	0.37	0.0008	0.0066	nd	0.0225	0.042
	0.078	1.46	0.38	0.0009	0.0065	nd	0.0265	0.038
Oxygen	0.088	1.54	0.43	0.0011	0.011	nd	0.0059	0.022
	0.085	1.66	0.42	0.0007	0.0096	nd	0.0070	0.025
	0.079	1.61	0.40	0.0005	0.0077	nd	0.0079	0.032

Notes: all welds contained 0.003–0.008%S, 0.005–0.008%P, ~0.03%Cr, Ni and Cu, 0.005%Mo, <5 ppm Al, Nb and V, unless otherwise stated.
N_t – total nitrogen, nd – not determined.

with C, Mn, Si and a range of microalloying elements. Tensile specimens from 3 beads/layer ISO 2560:1973 test pieces were tested at ambient temperature and at 250 °C, i.e. within the strain ageing regime, where increased strength and reduced ductility are expected in steel subject to strain ageing. The results were then compared. The compositions of the weld metals tested are listed in Table 12.8; subsequent determinations of free nitrogen on selected welds are detailed in Table 12.12.

Table 12.9 Tensile test results on weld metal for dynamic strain ageing tests

Series*	20 °C				250 °C				Difference		
	YS, N/mm²	TS, N/mm²	EI, %	RA, %	YS, N/mm²	TS, N/mm²	EI, %	RA, %	TS, N/mm²	RA, %	P_{sa}
0.045 C	490	520	29	79	430	660	22	55	140	−24	140
0.12 C	530	590	24	76	470	730	23	50	140	−26	140
0.5 Mn	440	490	25	78	400	620	23	57	130	−21	130
1.0 Mn	490	520	27	76	430	650	22	59	120	−16	120
1.4 Mn†	490	550	29	77	450	690	24	54	140	−23	140
1.8 Mn	560	600	28	74	530	730	20	53	120	−21	130
0.2 Si	510	540	27	75	420	680	23	57	140	−18	130
0.7 Si	540	600	26	73	490	720	23	55	120	−18	120
1.1 Si	590	640	27	72	520	720	20	56	85	−17	90
16 Al	510	570	nd	77	440	690	20	53	120	−24	130
53 Al	520	570	29	79	460	700	23	55	120	−24	130
85 Al	520	570	26	78	450	670	20	58	100	−19	110
130 Al	530	570	27	77	450	680	25	60	110	−17	110
0 Ti	460	540	22	78	410	660	21	59	120	−19	120
30 Ti	480	550	28	77	460	680	24	56	130	−22	130
120 Ti	510	570	26	78	470	680	22	53	110	−25	130
160 Ti	510	560	27	76	490	680	22	53	110	−23	120
35 B	520	560	nd	74	460	670	21	56	110	−18	120
54 B	500	540	27	74	440	650	nd	55	110	−18	110
69 B	470	550	30	73	420	640	21	65	90	−9	80
0 B	490	540	26	77	440	670	19	48	130	−29	140
28 B	520	550	27	77	440	660	19	59	110	−18	110
46 B	500	510	25	74	430	620	20	59	110	−15	110
63 B	480	500	28	70	390	560	19	55	60	−15	70
170 N	520	580	25	74	470	730	nd	44	160	−31	150
220 N	550	600	26	73	520	760	nd	39	160	−34	160
260 N	570	620	26	73	530	790	nd	28	170	−45	170
220 O	530	590	24	77	490	710	16	52	130	−25	130
250 O	540	600	28	76	500	710	24	57	110	−19	110
320 O	520	590	22	75	470	680	22	61	90	−14	100

Notes: El – elongation; RA – reduction of area; TS – tensile strength; YS – yield strength.
Difference is (value at 250 °C) minus (value at ambient temperature).
* coded as % for C, Mn & Si, ppm for others.
† baseline composition.

The results of preliminary tensile tests are given in Table 12.9, from which it can be seen that in tests at 250 °C, tensile strength and reduction of area were more influenced by dynamic strain ageing than were yield strength and elongation. A relationship was deduced between the change in RA (Δ_{RA}) and the change in tensile strength (Δ_{TS}):

$$\Delta_{TS} = 188\left(1 - e^{\Delta_{RA}/18}\right) \qquad [12.1]$$

Table 12.10 Composition of commercial and modified commercial weld metal for dynamic strain ageing tests

Element, wt%	C	Mn	Si	Ni	Mo	Ti	B	N_t
Electrode								
1[1]	0.068	1.51	0.43	nd	nd	0.0072	0.0007	0.0073
2[2]	0.043	1.38	0.74	nd	nd	0.0087	0.0007	0.0073
3[3]	0.052	1.34	0.37	0.93	0.35	0.0061	<0.0003	0.0098
3 + B	0.052	1.37	0.37	0.92	0.37	0.0054	0.0011	0.0091
3 + B′	0.050	1.31	0.37	0.92	0.33	0.0057	0.0023	0.0097
3 + B″	0.050	1.35	0.38	0.91	0.34	0.0056	0.0061	0.0085

Notes: N_t – total nitrogen;
(1) also contained 0.008%S, 0.004%P, 0.0008%Al, 0.038%O;
(2) also contained 0.009%S, 0.016%P, 0.0011%Al, 0.039%O;
(3) No. 3 series also contained 0.005–0.006%S, 0.008–0.010%P, 0.03%Cr, Cu, 0.01%V, 0.0003–0.0004%Al, <0.005%Sn, 0.005–0.010%Co, 0.006–0.007%As, 0.041–0.043%O.
nd – not determined.

This has been used to obtain a single parameter, P_{sa}, for dynamic strain ageing embrittlement:

$$P_{sa} = \Delta_{TS} + 188\left(1 - e^{\Delta RA/18}\right) \qquad [12.2]$$

and the values of this parameter for each weld are included in Table 12.9.

Most welds had P_{sa} parameters between 120 and 140. The welds with high nitrogen contents show the highest parameter values, rising from 140 to 170 as the nitrogen content was increased from 100 (baseline composition) to 265 ppm. The lowest parameter values (P_{sa} values of 70 and 80) were for welds containing the highest level of boron in each boron series. Other welds showing values below 100 were the weld of high silicon content (P_{sa} parameter 90) and one of the oxygen series with 320 ppm O. Silicon in amounts of 1% and more is not a desirable addition to MMA weld metals; the oxygen-series weld appears to be anomalous as welds of both higher (380 ppm – the baseline weld in the manganese series) and lower (250 ppm) oxygen content showed higher values of the parameter. These two approaches were, therefore, not followed up.

To explore in greater detail the possibility of improving resistance to dynamic strain ageing by adding boron, tests were made on commercial and modified commercial electrodes with boron added. Their weld metal compositions are given in Table 12.10 and tensile properties at ambient temperature and 250 °C in Table 12.11. Neither of

Table 12.11 Tensile test results for commercial and modified commercial weld metal from dynamic strain ageing tests

Elect.	20 °C				250 °C				Difference		
	YS, N/mm^2	TS, N/mm^2	El, %	RA, %	YS, N/mm^2	TS, N/mm^2	El, %	RA, %	TS, N/mm^2	RA, %	P_{sa}
1	520	580	29	75	460	690	nd	54	120	−21	120
2	500	550	29	78	460	660	20	62	110	−17	110
3	nd	650	23	73	nd	740	nd	60	90	−13	90
3 + B	nd	650	25	72	nd	750	nd	58	100	−14	100
3 + B′	nd	640	27	73	nd	760	nd	54	120	−19	120
3 + B″	nd	620	22	70	nd	710	nd	53	90	−16	90

Notes: El – elongation; RA – reduction of area; TS – tensile strength; YS – yield strength.
Difference is (value at 250 °C) minus (value at ambient temperature).
nd – not determined, specimen fractured outside gauge length.

the two commercial C–Mn electrodes (coded 1 and 2) gave particularly low values of the P_{sa} parameter, although electrode 2, with 0.7%Si, gave a similar value of 110 to the 0.7%Si experimental electrode (P_{sa} 120). The high strength Mn–Ni–Mo electrode, coded 3, both with and without boron, gave relatively low P_{sa} values of 90–120. Both the lowest and highest boron contents gave parameter values at the minimum, although this was appreciably above the values of 70 and 80 for the experimental boron-containing electrodes.

As-welded Charpy tests on the second boron series and the Mn–Ni–Mo electrodes (coded 3), both with and without boron, showed that increasing the boron content above about 30 ppm gave a sharp drop in toughness with a 10–20 °C increase in transition temperatures (Fig. 12.12). It seemed likely that any beneficial influence of boron on reduction of strain ageing embrittlement, as assessed by Charpy toughness testing, would be counterbalanced by starting from a base material of inherently lower toughness, so that the final result would be little different.

One likely reason for the ability of boron additions to reduce the amount of strain ageing embrittlement is that boron reduces the amount of free nitrogen (Ref. 4, Ch. 7). The present results, together with those of some tests carried out on submerged-arc weld metal, generally support this idea, as shown in Table 12.12. The submerged-arc results show a clear fall in free nitrogen content as boron was increased. The MMA results are apparently contradictory but when the two welds of similar total nitrogen are compared, the results fall into line; weld N21 of low total nitrogen content gave a correspondingly low free nitrogen value.

Table 12.12 Detailed nitrogen analyses of welds of varying boron content

Weld identity	Welding process	Boron, ppm	Nitrogen, ppm		
			Total	Residual	Free, by difference
N86	MMA	0.2	137	83	56
N21		1.9	32	26	5
N22		2.1	128	93	28
S75	Submerged-arc	0.8	104	77	34
S66		3.5	88	65	21
S67		5.4	98	82	16
S68		6.9	94	86	7

Note: total nitrogen values are means of two independent determinations, residual and free are single (duplicate) values.

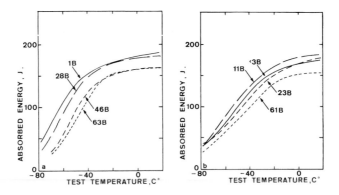

12.12 Charpy transition curves for welds with added boron (level given in ppm): *a)* Second series (Tables 12.8 and 12.9); *b)* Based on electrode 3 (Tables 12.10 and and 12.11).

Reduction of weld nitrogen level

One possible way to reduce the damaging effects of strain ageing would be to reduce the nitrogen content of the weld. Of the two main sources of nitrogen, the arc atmosphere and the electrode, the latter can only be controlled by restricting the nitrogen content of the core wire and any alloying material (particularly ferro-alloys) in the coating. In Chapter 11 (Table 11.1) it was shown that additions of magnesium to reduce the oxygen content from 0.041 to 0.024% also reduced nitrogen from 80–100 ppm to 50–60 ppm, although this is

Table 12.13 Composition of 1.4%Mn weld metal deposited from electrodes with additions of zinc

Element, wt%

C	Mn	Si	S	P	N	O	Zn
0.069	1.45	0.27	0.010	0.010	0.0094	0.043	0.0020
0.076	1.48	0.31	0.007	0.009	0.0075	0.043	0.0082
0.077	1.43	0.26	0.007	0.013	0.0080	0.041	0.0144
0.076	1.39	0.26	0.008	0.010	0.0075	0.040	0.0262

Note: welds also contained ~0.03%Cr, Ni, Cu, 0.005%Mo, ~27 ppm Ti, <10 ppm V, ≤5 ppm Al.

Table 12.14 Mechanical properties, as-welded, of 1.4%Mn welds, with zinc additions to coating

Zn, ppm	Yield strength, N/mm²	Tensile strength, N/mm²	Elongation, %	Reduction of area, %	Charpy transition, °C	
					100 J	28 J
20	480	540	28	77	−65	−90
82	500	540	27	76	−65	−94
144	480	540	26	78	−65	−88
262	470	530	29	77	−65	−89

appreciably above the nitrogen content of the core wire itself (20–30 ppm, Chapter 2).

With sufficient magnesium to give 0.031%O and an addition of 220 ppm Al a level of 40 ppm N was attained. However, on its own, aluminium does not greatly reduce nitrogen. In Table 12.3 it can be seen that addition of as much as 610 ppm Al only reduced nitrogen from 70 to 50 ppm, although greater reductions were achieved from higher nitrogen levels with similar additions, i.e. 150 to 120 ppm N in one instance and 240 to 200 ppm N in the other.

A further, hitherto unreported, experiment was made to assess the effect of the volatile metal zinc which, unlike magnesium and aluminium, has no great affinity for nitrogen. The composition of 1.4%Mn–30 ppm Ti deposits with additions of zinc powder to the coating in Table 12.13 reveals a small reduction in nitrogen from 94 to 75 ppm (similar to the maximum effect seen with aluminium) with a weld containing 260 ppm Zn. No significant changes were encountered in the other elements, although manganese, sulphur and oxygen were marginally reduced.

Mechanical properties (Table 12.14) were unchanged by the additions of zinc.

Discussion

It was concluded that, although additions of volatile elements such as magnesium and zinc are capable of reducing nitrogen, none of the tests succeeded in reaching the nitrogen level of the core wire, let alone lower levels. Furthermore, the time scale involved in welding is such that combination of nitrogen with nitride-forming elements is unlikely to go to completion in the as-welded condition. The best guard against its embrittling effect is to use weld metals of good as-welded toughness to allow for any subsequent reduction in toughness.

References

1 Koçak M, Petrovski BI and Evans GM: 'Influence of titanium and nitrogen on the fracture properties of weld metals'. Proc int OMEA conf, Houston, 1994.
2 Koçak M, Petrovski BI and Evans GM: 'Effect of N, Ti and strain ageing on the toughness of weld metals'. Proc conf 'Eurojoin 2', Florence, 1994.
3 Evans GM: 'The effect of nitrogen on C–Mn steel welds containing titanium and boron'. Proc 76th AWS Annual Convention, 1995; IIW Doc II–1288–96.
4 Bailey N and Evans GM: 'Strain ageing embrittlement of repair weld metals for service up to 350 °C; tests by Oerlikon'. TWI Report 5555/33/89, 1989.

Part IV

Microalloying of C–Mn
steel weld metals

13

Single microalloying elements (with and without titanium)

Small quantities of certain elements can profoundly affect the microstructure and properties of C–Mn weld metals. Such elements – often referred to collectively as microalloying elements – can be introduced, perhaps unwittingly, from parent steel, electrode core wire or flux ingredients or they may be added deliberately. In the tests described in Part I, sufficient amounts of microalloying elements were present to influence weld metal properties significantly; the present tests were made using baseline weld metals containing less than 5 ppm of any of them, so that any effects could be unambiguously revealed.

The elements involved are titanium, aluminium, vanadium, niobium and boron. Titanium itself was studied in Chapter 10 and its relation to oxygen and nitrogen in Chapters 11 and 12. This chapter covers addition of the elements aluminium, vanadium, niobium and boron made singly, both in the absence and in the presence of small quantities (~30 ppm) of titanium and usually at several levels of manganese. Additions of more than one microalloying element, including other microalloying elements with more than 30 ppm Ti, are dealt with in the next chapter.

Aluminium (Al)

Aluminium is not generally added to MMA electrodes. However, the tendency to replace rimming steel core wires (which contain no aluminium) by wires originating from electric steelmaking, using mainly scrap charges and continuous casting (a technology actually requiring aluminium-treated steels), is increasing the chance of aluminium entering the deposit. Aluminium is a deliberate addition to some self-shielded flux-cored electrodes (and also some unshielded solid wires). In high dilution welding it can also be picked up from aluminium-treated parent steels, particularly in submerged-arc welding, where

Table 13.1 Composition of 1.4% and 0.6%Mn weld metal without and with added titanium and having varying aluminium content

Element, wt%	C	Mn	Si	S	P	Al	Ti	N	O	Al/O ratio
Electrode code										
C₀ (no Ti)	0.074	1.49	0.29	0.006	0.013	0.0006	0.0005	0.007	0.047	0.01
	0.081	1.43	0.29	0.006	0.013	0.0060	0.0005	0.005	0.046	0.13
	0.076	1.36	0.34	0.005	0.011	0.015	0.0005	0.005	0.048	0.31
	0.078	1.37	0.39	0.005	0.011	0.025	0.0005	0.005	0.045	0.56
	0.081	1.40	0.44	0.004	0.011	0.042	0.0004	0.004	0.044	0.95
	0.080	1.42	0.48	0.004	0.006	0.066	0.0005	0.005	0.044	1.50
A₀ (no Ti)	0.063	0.64	0.36	0.007	0.007	0.0005	0.0005	0.006	0.048	0.01
	0.076	0.68	0.25	0.007	0.010	0.0082	0.0001	0.005	0.049	0.17
	0.079	0.71	0.37	0.007	0.007	0.020	0.0003	0.005	0.046	0.23
	0.080	0.74	0.42	0.007	0.007	0.030	0.0003	0.005	0.046	0.65
	0.076	0.67	0.44	0.005	0.006	0.044	0.0002	0.005	0.044	1.00
	0.075	0.68	0.68	0.005	0.005	0.052	0.0003	0.004	0.035	1.50
A	0.068	0.58	0.27	0.009	0.009	<0.0005	0.0036	0.008	0.051	0.01
	0.074	0.72	0.33	0.007	0.007	0.0064	0.0044	0.006	0.050	0.13
	0.078	0.72	0.39	0.006	0.007	0.020	0.0046	0.005	0.048	0.41
	0.075	0.67	0.41	0.005	0.006	0.028	0.0040	0.006	0.051	0.55
	0.077	0.70	0.45	0.005	0.007	0.055	0.0050	0.005	0.043	1.27
	0.076	0.71	0.49	0.004	0.008	0.082	0.0055	0.006	0.038	2.15

Notes: see Table 12.3 for compositions of deposits with 35 ppm Ti and 1.4%Mn (series C).
All welds contained <5 ppm B, Nb, V, ~0.03%Cr, Cu, Ni and ~0.005%Mo.
The Al/O ratio for Al_2O_3 is 1.124.

the element aluminium can also be introduced from alumina in semi-basic and basic fluxes and where it is known (Ref. 8, Ch. 3 and Ref. 6, Ch. 10) that excessive aluminium can prevent formation of acicular ferrite and hence give poor toughness. In steels generally, aluminium is the most common way of fixing nitrogen and thus avoiding strain-ageing embrittlement, and this aspect was discussed in Chapters 7 and 12.

The composition and mechanical properties of 1.4%Mn welds with a small addition of titanium (~35 ppm) and varying aluminium contents[1] were given in Tables 12.3 and 12.4, respectively, where they formed the basis of an investigation into combined effects of nitrogen and aluminium. Several of these welds were also used in Chapter 11 for an examination of effects of oxygen and titanium. An additional series with no added titanium (Ref. 3, Ch. 10) was examined and the composition of these welds (C_o) is detailed in Table 13.1. Also included are the results for deposits containing 0.6%Mn (A and A_o).

Standard test methods were used; aluminium was added as aluminium powder to the electrode coatings; details of recoveries are given in Chapter 2. Only the series with 1.4%Mn and titanium was tested after stress relief. Analyses of the welds tested (Tables 12.3 and 13.1) reveal that, although the contents of carbon, manganese and most other elements were sensibly constant, silicon increased when aluminium content exceeded about 100 ppm, whilst nitrogen and sulphur decreased.

Metallography

In the welds containing 1.4%Mn, no difference in weld structures on a macro-scale was apparent. The microstructure of the top beads of deposits with titanium was of the acicular ferrite type, whereas without added titanium it was a mixture of primary ferrite and ferrite with aligned second phase. Adding aluminium had contrary effects on the two series (Fig. 13.1). Without added titanium aluminium increased acicular ferrite at the expense of ferrite with second phase, but with titanium (Fig. 13.1(b)), the reverse occurred.

Although the content of acicular ferrite doubled when aluminium was added to the titanium-free series (Fig. 13.1(a)), the increase was only from ~10 to ~20% over the whole range of aluminium so that the amount of ferrite with second phase fell from about 55 to 40%.

The changes in the coarse reheated regions mirrored those in the as-deposited microstructure. In fine grained regions, aluminium initially refined and then (above 30 ppm Al) coarsened this region. As

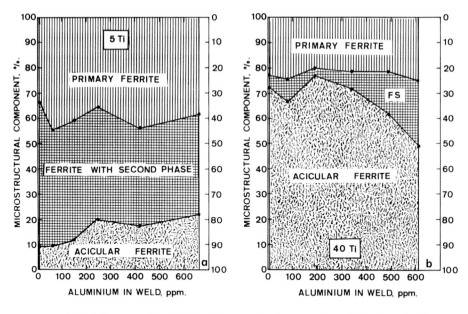

13.1 Influence of weld aluminium content on as-deposited microstructure of deposits with: a) No added titanium; b) 35 ppm Ti.

with other alloying additions, aluminium also changed the second phase from degenerate pearlite type to a more duplex structure with M/A constituent. A discussion of the non-metallic inclusion population is included in Chapter 17.

Mechanical properties

Aluminium had little effect on hardness of welds with ~35 ppm titanium and averaged 210 HV5 ±10 HV.

Mechanical properties of 1.4%Mn welds with no added titanium (C_0) are given in Table 13.2; results of the welds with titanium are in Table 12.4. Also incorporated in Table 13.2 are equivalent data for welds containing 0.6%Mn. As-welded strength values for the C_0 series are plotted against aluminium content in Fig. 13.2. Although strength values tended to increase slightly in all tests, this was probably a result of an increase in silicon content as aluminium content was increased.

Without titanium, toughness was poor in welds with 1.4%Mn but, after the first small aluminium addition of 60 ppm, it improved slightly and progressively with added aluminium (Fig. 13.3).

Table 13.2 As-welded mechanical properties of 1.4%Mn welds without added titanium and 0.6%Mn welds having varying aluminium content

Elect code	Al, ppm	Yield strength, N/mm²	Tensile strength, N/mm²	Elongation, %	Reduction of area, %	Charpy transition, °C	
						100 J	28 J
C₀	6	430	530	30	78	−19	−41
(no	60	440	540	28	73	−15	−39
Ti)	150	440	530	25	79	−20	−58
	250	440	540	25	79	−56	−62
	420	470	540	24	77	−30	−62
	660	470	570	26	79	−29	−76
A₀	5	410	480	30	77	−53	−78
(no	82	390	480	26	77	−64	−78
Ti)	200	400	480	29	79	−60	−81
	300	410	490	27	76	−57	−89
	440	420	490	27	76	−57	−89
	520	430	500	27	76	−55	−85
A	<5	400	490	32	77	−48	−65
(no	64	430	500	33	77	−31	−52
Ti)	200	440	510	31	79	−39	−55
	280	440	520	35	78	−34	−49
	550	470	540	31	75	−50	−76
	820	450	550	31	73	−51	−73

Note: see Table 12.4 for properties of welds with 36 ppm Ti and 1.4%Mn (series C).

As-welded Charpy toughness was good in the Al-free welds with titanium. Some increase of the 100 J transition temperature was apparent (Fig. 13.3) with addition of up to 100 ppm Al and a further, larger increase was apparent at the highest aluminium content of 610 ppm Al, so this weld was only marginally tougher than the weld of similar aluminium content without added titanium. Charpy temperatures after stress relief parallelled the as-welded values and averaged 12 °C lower. It is noteworthy that the weld of highest aluminium content was the only one to contain sufficient aluminium to tie up all the available oxygen as Al_2O_3 and prevent other oxides forming, a feature which has been believed in submerged-arc welds to prevent nucleation of acicular ferrite when conditions appeared to be otherwise suitable (Ref. 8, Ch. 3). Aluminium had little influence on the toughness of welds with 0.6%Mn, but tended to be slightly beneficial – even in the presence of titanium.

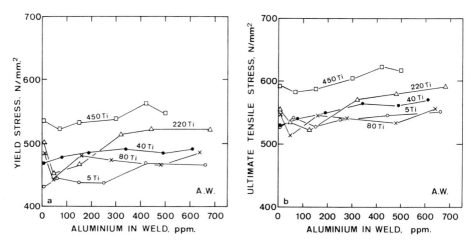

13.2 Influence of weld aluminium on: *a)* Yield strength; *b)* Tensile strength, of 1.4%Mn welds with different titanium contents.

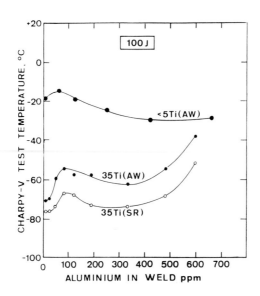

13.3 Influence of weld aluminium on 100 J Charpy transition temperature of 1.4%Mn welds.

Vanadium (Mn–V)

Although vanadium is used as an alloying element in certain high temperature and tool steels, it is a rare constituent of mild, C–Mn and low alloy steel weld metals – largely because it tends to give poor toughness after stress relief and also is influential in promoting reheat (stress relief) cracking in the HAZ of some steels. Small amounts can enter MMA welds as an impurity in rutile and coating materials or in iron powder. The present work examined the possible role of vanadium in smaller quantities (up to 0.1%) as a ferrite refiner, because vanadium has been found to be capable of nucleating acicular ferrite in the HAZ of some steels containing the element. Ferro-vanadium in the coating was used to add vanadium and typical recoveries are given in Chapter 2.

Two series of welds were made; the first, comprising welds with the four standard manganese levels, was made with a small addition of titanium and low levels of other micro-alloying elements.[2] The second, hitherto unreported series, with both 0.6 and 1.4%Mn (series A_o and C_o), had no added titanium; both series were tested as-welded and after stress relief heat treatment.

Weld analysis (Table 13.3), shows well balanced compositions, the only departure from the ideal being a slight tendency for titanium levels in the first series to be reduced as vanadium was increased.

Metallography

In the series of welds with a titanium addition, vanadium increased the proportion of acicular ferrite at the expense of the other two constituents (Fig. 13.4), particularly in welds of low manganese content. The increase in acicular ferrite continued to the highest vanadium level, even though the titanium content was slightly lower in these deposits. In addition to altering the proportions of the constituents, vanadium also refined the acicular ferrite in the microstructure as shown in Fig. 13.5.

Changes in coarse-grained reheated microstructure were similar to those in as-deposited weld metal. As with other alloying additions, vanadium gave the refined microstructure a more duplex appearance (Fig. 13.6), increasing the proportion of the M/A constituent. Stress relief at 580 °C led to precipitation of carbides (Fig. 13.7). Vanadium had little influence on the grain size of the fine-grained regions at low manganese levels but appreciably refined these regions when 1.4%Mn was present (Fig. 13.8).

Table 13.3 Composition of weld metal having varying vanadium and manganese content

Element, wt%	C	Mn	Si	S	P	V	Ti	N	O
Electrode code									
A	0.073	0.64	0.35	0.005	0.004	0.0003	0.0038	0.008	0.043
	0.071	0.63	0.36	0.007	0.006	0.021	0.0043	0.008	0.044
	0.071	0.64	0.40	0.008	0.006	0.044	0.0036	0.009	0.041
	0.074	0.63	0.36	0.008	0.005	0.060	0.0031	0.008	0.043
	0.072	0.64	0.36	0.008	0.006	0.082	0.0034	0.008	0.044
B	0.077	1.02	0.35	0.007	0.005	0.0004	0.0041	0.008	0.040
	0.076	1.01	0.34	0.007	0.007	0.022	0.0037	0.008	0.043
	0.078	1.04	0.31	0.007	0.006	0.044	0.0032	0.005	0.043
	0.076	1.04	0.33	0.007	0.006	0.061	0.0031	0.007	0.043
	0.077	1.05	0.32	0.008	0.006	0.085	0.0034	0.007	0.040
C	0.077	1.33	0.31	0.007	0.005	0.0004	0.0038	0.008	0.042
	0.072	1.22	0.33	0.008	0.006	0.019	0.0029	0.008	0.045
	0.077	1.36	0.26	0.007	0.007	0.043	0.0029	0.008	0.041
	0.078	1.35	0.30	0.007	0.007	0.060	0.0030	0.008	0.040
	0.076	1.36	0.26	0.007	0.006	0.100	0.0029	0.008	0.041
D	0.076	1.74	0.29	0.008	0.006	0.0004	0.0035	0.009	0.042
	0.077	1.73	0.26	0.007	0.008	0.024	0.0027	0.007	0.042
	0.077	1.72	0.26	0.007	0.008	0.041	0.0025	0.007	0.042
	0.076	1.72	0.25	0.005	0.008	0.060	0.0025	0.009	0.040
	0.080	1.74	0.26	0.007	0.008	0.083	0.0025	0.009	0.041
A$_0$ (no Ti)	0.064	0.65	0.36	0.007	0.007	0.0005	<0.0005	0.006	0.049
	0.066	0.67	0.34	0.008	0.008	0.0195	<0.0005	0.006	0.047
	0.064	0.68	0.31	0.007	0.006	0.042	<0.0005	0.007	0.048
	0.066	0.65	0.32	0.007	0.006	0.0605	<0.0005	0.008	0.049
	0.065	0.66	0.32	0.006	0.006	0.0815	<0.0005	0.006	0.046
C$_0$ (no Ti)	0.072	1.44	0.32	0.006	0.008	0.0006	<0.0005	0.008	0.044
	0.070	1.46	0.28	0.006	0.008	0.026	<0.0005	0.007	0.044
	0.070	1.47	0.29	0.007	0.008	0.042	<0.0005	0.008	0.044
	0.070	1.48	0.30	0.006	0.008	0.063	<0.0005	0.006	0.044
	0.066	1.44	0.28	0.006	0.008	0.0825	<0.0005	0.008	0.046

Notes: all welds contained <5 ppm Al, B, <6 ppm Nb, ~0.03% Cr, Cu, Ni and ~0.005% Mo

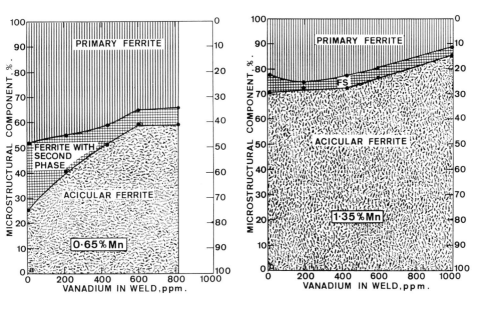

13.4 Influence of weld vanadium content on as-deposited microstructure of welds with: a) 0.6%Mn and 40 ppm Ti; b) 1.4%Mn and 40 ppm Ti.

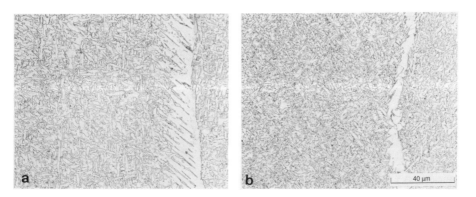

13.5 Microstructure of as-deposited regions of welds with 1.4%Mn and 40 ppm Ti with: a) 4 ppm V; b) 0.1%V.

Mechanical properties

When titanium was present, vanadium increased as-welded hardness linearly, except that the first addition of vanadium to the low manganese weld A gave a particularly large increase in hardness, as shown in Fig. 13.9.

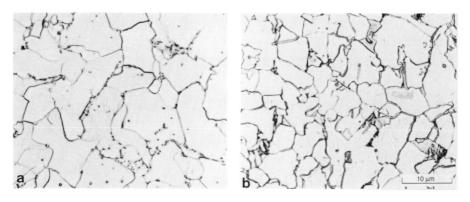

13.6 Microstructure of fine-grained regions of welds (as-welded) with 1.4%Mn and 40 ppm Ti with: *a*) 4 ppm V; *b*) 0.1%V.

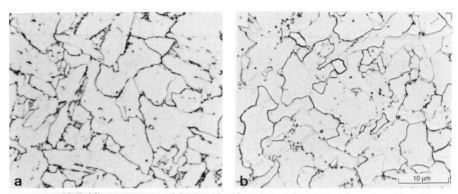

13.7 Microstructure of fine-grained regions of stress-relieved welds with 1.4%Mn and 40 ppm Ti with: *a*) 4 ppm V; *b*) 0.1%V.

Vanadium, as well as manganese, increased strength (Table 13.4), and was particularly effective as a strengthener in the PWHT condition (Table 13.4(*b*)), even at the smallest addition of ~200 ppm. Vanadium was a less effective strengthener in the absence of titanium in the as-welded condition (Table 13.4(*c*)), although this was no longer true after stress-relief.

Overall, vanadium minimised the drop in strength on stress relieving. Although little softening occurred in some of the deposits, no weld was actually stronger after PWHT than before. Figure 13.10 shows that the effect of vanadium was not linear, vanadium having a decreasing effect as more was added; the following regression equations (V and Mn both in wt%) were obtained for deposits with added titanium:

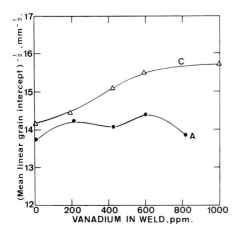

13.8 Influence of weld vanadium content on grain size of fine-grained regions with: *a)* 0.6%Mn (A); *b)* 1.4%Mn (C).

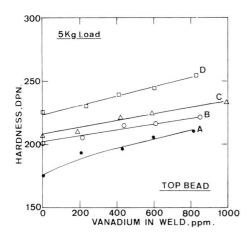

13.9 Influence of weld vanadium content on as-welded hardness with titanium at varying manganese levels.

$$YS_{aw} = 331 + 103\,Mn + 1575\,V - 3478\,V^2 \qquad [13.1]$$
$$TS_{aw} = 430 + 85\,Mn + 1162\,V - 2077\,V^2 \qquad [13.2]$$

$$YS_{sr} = 273 + 83\,Mn + 3613\,V - 19427\,V^2 \qquad [13.3]$$
$$TS_{sr} = 395 + 76\,Mn + 2425\,V - 10569\,V^2 \qquad [13.4]$$

There was no significant interaction between vanadium and manganese.

Table 13.4 Mechanical properties of welds with varying vanadium and manganese content: a) As-welded deposits with ~35 ppm Ti

Mn, %	V, ppm	Yield strength, N/mm^2	Tensile strength, N/mm^2	Elongation, %	Reduction of area, %	Charpy transition, °C	
						100 J	28 J
0.6, A	3	400	490	33	78	−45	−70
	210	440	510	29	77	−43	−62
	440	480	550	28	78	−43	−63
	600	490	550	26	78	−42	−60
	820	490	560	25	77	−40	−59
1.0, B	4	440	520	31	77	−60	−84
	220	460	530	25	80	−69	−86
	440	500	570	25	77	−63	−86
	610	510	580	26	77	−59	−85
	850	540	590	25	74	−60	−78
1.4, C	4	460	540	27	76	−68	−89
	190	490	540	30	78	−66	−87
	420	540	590	24	78	−70	−94
	600	520	590	23	76	−70	−89
	1000	600	640	21	74	−55	−83
1.8, D	4	500	570	29	77	−62	−90
	240	540	610	23	77	−62	−88
	430	580	620	22	75	−67	−90
	600	590	650	22	75	−64	−90
	830	640	680	20	72	−61	−93

Charpy tests, also included in Table 13.4, are summarised in terms of 100 J transition temperature for welds with 40 ppm Ti in Fig. 13.11. As-welded, vanadium had little effect on Charpy toughness of deposits with titanium, the results for each manganese series lying within a scatter band no more than 10 J wide. However, vanadium improved the toughness of 1.4%Mn welds with no added titanium (Table 13.4(c)), although not to the level of the corresponding welds with titanium. However, vanadium was slightly detrimental to deposits with 0.6%Mn without added titanium, although these welds were all slightly tougher than 0.6%Mn welds with added titanium. The improvement of the titanium-free 1.4%Mn deposits was evidently a result of vanadium promoting acicular ferrite and refining the microstructure of such welds.

In the stress-relieved condition (Fig. 13.11(b)), vanadium was particularly detrimental, even the smallest addition of ~200 ppm perceptibly increased transition temperature, except for the 1.4%Mn weld without added titanium, where formation of acicular ferrite appeared to have offset the influence of the first few hundred ppm V.

Table 13.4 (cont.) b) Deposits with ~35 ppm Ti stress-relieved at 580 °C

Mn, %	V, ppm	Yield strength, N/mm²	Tensile strength, N/mm²	Elongation, %	Reduction of area, %	Charpy transition, °C 100 J	28 J
0.6, A	3	330	450	36	80	−55	−72
	210	400	480	30	76	−47	−61
	440	450	530	nd	76	−30	−50
	600	460	550	26	76	−28	−47
	820	480	580	25	76	−22	−44
1.0, B	4	330	460	33	80	−80	−95
	220	430	520	30	79	−67	−86
	440	480	550	27	78	−51	−79
	610	490	570	27	77	−45	−73
	850	540	610	25	75	−40	−65
1.4, C	4	370	500	31	76	−81	−97
	190	460	530	28	78	−67	−81
	420	530	590	26	76	−62	−85
	600	530	600	26	75	−60	−85
	1000	570	640	23	73	−36	−60
1.8, D	4	410	520	29	78	−71	−99
	240	520	600	26	75	−59	−82
	430	540	630	22	74	−55	−79
	600	520	600	24	74	−44	−72
	830	570	650	24	74	−32	−60

Table 13.4 (cont.) c) Deposits with no added titanium

Mn, %	V, ppm	Yield strength, N/mm²	Tensile strength, N/mm²	Elongation, %	Reduction of area, %	Charpy transition, °C 100 J	28 J
0.6, A_0 (AW)	5	400	480	30	77	−53	−78
	190	420	490	29	77	−56	−78
	420	440	500	27	78	−53	−75
	600	470	520	25	74	−47	−74
	810	490	540	26	76	−45	−69
1.4, C_0 (AW)	5	460	550	27	77	−24	−42
	260	480	550	24	78	−45	−72
	420	480	560	24	76	−48	−82
	630	540	580	24	76	−57	−84
	820	540	590	21	76	−44	−77
0.6, A_0 (SR)	5	340	440	35	75	−58	−73
	190	380	470	28	77	−47	−68
	420	420	520	27	79	−35	−50
	600	440	520	27	74	−27	−43
	810	440	540	26	75	−16	−38
1.4, C_0 (SR)	6	360	490	29	78	−61	−97
	260	460	550	26	78	−60	−91
	420	500	580	23	75	−53	−88
	630	520	600	23	74	−28	−62
	820	520	600	22	74	−10	−57

Note: AW – as-welded; SR – stress relieved at 580 °C.

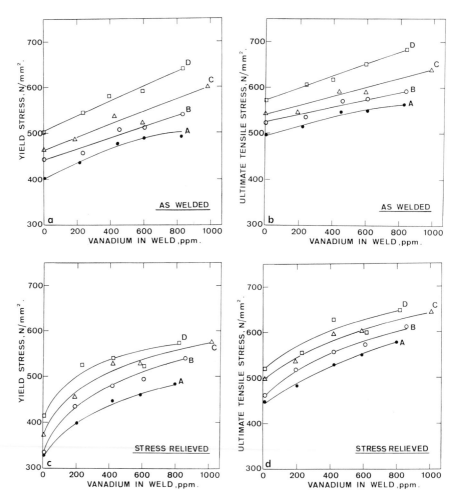

13.10 Influence of weld vanadium content of C–Mn deposits on: *a*) Yield strength, as-welded; *b*) Tensile strength, as-welded; *c*) Yield strength, stress relieved at 580 °C; *d*) Tensile strength, stress relieved at 580 °C.

This embrittlement by vanadium after stress relief is particularly well illustrated in Fig. 13.12, which shows the shift of transition temperature on stress relieving deposits with titanium. Overall, the optimum content of ~1.4%Mn welds with ~40 ppm Ti was maintained in both test conditions.

Discussion

Vanadium is an element which tends to form strengthening and embrittling carbides on heat treatment; this clearly explains the resistance

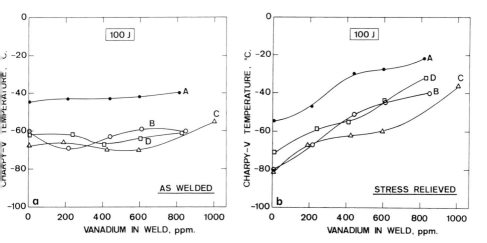

13.11 Influence of weld vanadium content on 100 J Charpy transition temperature of C–Mn deposits with 40 ppm Ti and varying manganese content: *a*) As-welded; *b*) Stress relieved at 580 °C.

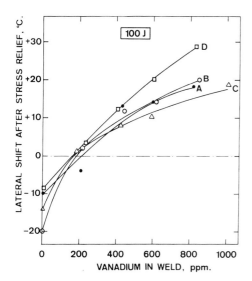

13.12 Influence of weld vanadium content on the shift in Charpy transition temperature on stress relief for deposits with varying manganese content and 40 ppm Ti.

of the strength properties to, and the drop in toughness on, PWHT. Formation of embrittling carbides is also consistent with the puzzling behaviour of deposits containing titanium in the as-welded condition. Because vanadium increased the amount and refined the structure of acicular ferrite, it would be expected to improve toughness (as it does in the case of titanium-free welds). However, formation of embrittling precipitates (consistent with the increase in hardness) in underlying weld beads which have been subjected to re-heating during welding could explain this apparently inconsistent behaviour.

The interdependence of vanadium and titanium is obviously a major cause of sometimes contradictory results in the literature where vanadium appears to be beneficial in some circumstances and harmful in others. Nevertheless, the shifts in Charpy temperature seen in Fig. 13.12 give a rational basis for limiting the vanadium content of mild steel and C–Mn weld metals to a maximum of about 0.02% (200 ppm), even where a fabrication is to be used in the as-welded condition, as it is always possible that some future modification or repair will necessitate PWHT.

Niobium (Mn–Nb)

Niobium is a common addition to structural steels and, although it is not deliberately added to MMA electrodes, it can frequently enter weld metal in small quantities, either diluted from the parent steel, from impurities in rutile or other coating materials or from the core wire. In the HAZ, niobium is known to give toughness problems in welds with high heat inputs and in combination with vanadium after PWHT.

Two test series were examined; the first[3] involved weld metals of four manganese levels (A–D) with a small deliberate addition of about 35 ppm Ti but with other micro-alloying elements at very low levels. The other, hitherto unreported series, used 0.6 and 1.4%Mn (A$_0$ and C$_0$) with no added titanium. Niobium was added up to 0.1% as ferro-niobium and recoveries are detailed in Chapter 2. Tests were carried out as-welded and after stress relief heat treatment at 580 °C. Weld compositions, given in Table 13.5, show them to be well balanced.

Metallography

Examination was restricted to the welds with added titanium, in which niobium decreased the amount of acicular ferrite and increased that of ferrite with second phase in the as-deposited microstructure (Fig. 13.13). With both 0.6 and 1.4%Mn, the first addition of niobium sharply reduced the content of acicular ferrite by about 10%. At the higher manganese level, a further drop in acicular ferrite occurred with the highest niobium addition (Fig. 13.14).

The effects of niobium were similar in high temperature, coarse reheated regions. As with several other alloying elements, niobium

Table 13.5 Composition of weld metal having varying niobium and manganese content

Element, wt%	C	Mn	Si	S	P	Nb	Ti	N	O
Electrode code									
A	0.073	0.64	0.35	0.005	0.004	<0.0005	0.0038	0.008	0.043
	0.073	0.64	0.34	0.007	0.005	0.012	0.0041	0.008	0.043
	0.075	0.62	0.33	0.007	0.005	0.022	0.0042	0.008	0.043
	0.070	0.64	0.33	0.007	0.004	0.045	0.0042	0.008	0.041
	0.074	0.64	0.33	0.008	0.006	0.088	0.0039	0.011	0.046
B	0.077	1.02	0.35	0.007	0.005	<0.0005	0.0041	0.008	0.040
	0.078	1.02	0.35	0.007	0.005	0.012	0.0036	0.008	0.037
	0.074	1.00	0.34	0.007	0.006	0.026	0.0037	0.009	0.042
	0.074	0.99	0.33	0.008	0.006	0.046	0.0037	0.009	0.039
	0.075	1.02	0.34	0.006	0.006	0.098	0.0037	0.008	0.042
C	0.077	1.33	0.31	0.007	0.005	0.0006	0.0038	0.008	0.042
	0.079	1.37	0.33	0.007	0.005	0.010	0.0034	0.010	0.042
	0.076	1.36	0.30	0.007	0.006	0.024	0.0033	0.008	0.037
	0.077	1.39	0.34	0.008	0.006	0.046	0.0032	0.010	0.041
	0.076	1.36	0.30	0.007	0.006	0.094	0.0033	0.009	0.043
D	0.076	1.74	0.29	0.008	0.006	<0.0006	0.0035	0.009	0.042
	0.077	1.77	0.32	0.007	0.006	0.012	0.0031	0.009	0.042
	0.078	1.76	0.30	0.007	0.007	0.023	0.0032	0.011	0.043
	0.077	1.77	0.29	0.008	0.009	0.045	0.0031	0.010	0.043
	0.079	1.79	0.29	0.006	0.007	0.090	0.0033	0.010	0.041
A_0	0.064	0.65	0.36	0.007	0.006	0.0005	<0.0005	0.006	0.049
	0.062	0.65	0.34	0.006	0.008	0.010	<0.0005	0.008	0.050
	0.062	0.64	0.32	0.007	0.008	0.022	<0.0005	0.007	0.050
	0.064	0.65	0.31	0.008	0.008	0.044	<0.0005	0.008	0.050
	0.062	0.66	0.32	0.008	0.010	0.092	<0.0005	0.007	0.050
C_0	0.072	1.54	0.32	0.006	0.008	0.0005	<0.0005	0.008	0.044
	0.071	1.54	0.32	0.007	0.009	0.012	<0.0005	0.008	0.044
	0.071	1.54	0.30	0.007	0.009	0.024	<0.0005	0.009	0.045
	0.070	1.54	0.30	0.007	0.010	0.046	<0.0005	0.006	0.046
	0.070	1.54	0.30	0.007	0.010	0.094	<0.0005	0.007	0.045

Note: all welds contained ≤7 ppm V, <5 ppm B, Al, ~0.03%Cr, Cu, Ni and ~0.005%Mo.

(up to 450 ppm) refined the fine-grained weld metal regions (Fig. 13.15) and gave them a more duplex appearance, the M/A constituent being prominent. Stress relief induced carbide precipitation in these regions.

Mechanical properties

In the as-welded condition, niobium increased the hardness of deposits containing titanium (Fig. 13.16), the increase being most

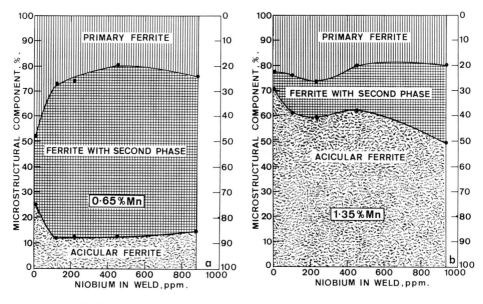

13.13 Influence of weld niobium content on as-deposited microstructure of welds with 40 ppm Ti and: *a*) 0.6%Mn; *b*) 1.4%Mn.

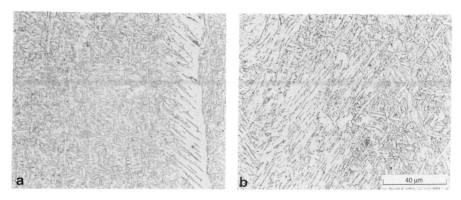

13.14 Microstructures of as-deposited regions of welds with 1.4%Mn and 40 ppm Ti with: *a*) 6 ppm Nb; *b*) 0.094%Nb.

apparent with the first addition of 100 ppm Nb, particularly at the lowest manganese level.

Niobium led to increases in strength and decreases in ductility (Table 13.6). Strength values showed a parabolic relationship (Fig 13.17), small initial niobium additions having a relatively large effect

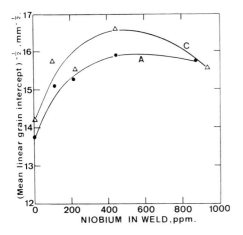

13.15 Influence of weld niobium content on grain size of fine-grained regions of welds with titanium and 0.6%Mn (A) and 1.4%Mn (C).

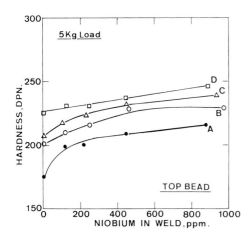

13.16 Influence of weld niobium content on as-welded hardness of welds with ~35 ppm Ti at varying manganese levels.

– particularly in the PWHT condition. Regression equations developed for the welds with 35 ppm Ti showed no significant interaction between the two strengthening elements niobium and manganese (both in wt%) but a stronger influence of niobium after stress relief than as-welded:

$$YS_{aw} = 354 + 85 \text{ Mn} + 3319 \text{ Nb} - 18062 \text{ Nb}^2 \qquad [13.5]$$
$$TS_{aw} = 449 + 74 \text{ Mn} + 2630 \text{ Nb} - 13498 \text{ Nb}^2 \qquad [13.6]$$

Table 13.6 Mechanical properties of welds with varying niobium and manganese content: *a*) As-welded deposits with ~35 ppm Ti

Electrode code	Nb, ppm	Yield Strength, N/mm^2	Tensile strength, N/mm^2	Elongation, %	Reduction of area, %	Charpy transition, °C	
						100 J	28 J
A	<5	400	490	33	78	−45	−70
	120	450	530	28	76	−34	−56
	220	470	550	27	81	−22	−45
	450	520	580	25	76	−7	−38
	880	550	610	24	71	13	−16
B	<5	440	520	31	77	−60	−84
	120	480	560	25	75	−42	−71
	260	510	590	28	76	−38	−58
	460	550	620	23	74	−11	−40
	980	620	660	20	73	14	−23
C	<6	460	540	27	76	−68	−89
	100	500	590	26	78	−55	−82
	240	540	600	26	76	−45	−73
	460	590	640	24	76	−23	−53
	940	610	680	21	72	−1	−33
D	<5	500	570	29	77	−62	−90
	120	540	610	22	76	−54	−86
	230	580	640	23	74	−42	−68
	450	610	660	22	74	−26	−62
	900	650	710	19	71	−6	−45

Table 13.6 (*cont.*) *b*) Deposits with ~35 ppm Ti stress-relieved at 580 °C

Electrode code	Nb, ppm	Yield Strength, N/mm^2	Tensile strength, N/mm^2	Elongation, %	Reduction of area, %	Charpy transition, °C	
						100 J	28 J
A	<5	330	450	26	80	−55	−72
	120	420	500	25	79	−38	−53
	220	450	540	24	78	−27	−38
	450	480	580	23	75	−5	−29
	880	540	620	23	72	18	−11
B	<5	330	460	33	80	−80	−95
	120	450	540	30	77	−55	−79
	260	490	580	25	76	−40	−62
	460	540	620	22	75	−20	−32
	980	600	680	20	71	8	−20
C	<6	370	500	31	76	−81	−97
	100	500	580	25	76	−61	−79
	240	540	610	25	76	−40	−62
	460	580	650	23	72	−27	−54
	940	610	690	19	71	7	−18
D	<5	410	520	29	78	−71	−99
	120	520	600	24	74	−48	−81
	230	550	630	22	77	−31	−65
	450	580	660	22	72	−15	−54
	900	640	720	17	69	4	−35

Table 13.6 (cont.) c) Deposits with no added titanium

Electrode code and state	Nb, ppm	Yield strength, N/mm²	Tensile strength, N/mm²	Elongation, %	Reduction of area, %	Charpy transition, °C	
Series						100 J	28 J
AW	5	410	480	30	77	−53	−78
A₀	90	440	520	27	77	−49	−67
	210	470	530	24	76	−44	−67
	430	520	570	24	75	−21	−48
	900	570	620	23	74	−2	−34
AW	5	460	550	27	77	−24	−42
C₀	120	510	590	26	76	−35	−68
	240	560	610	22	75	−33	−69
	460	580	640	21	75	−19	−59
	960	660	700	19	72	−2	−32
SR	5	330	440	35	75	−58	−73
A₀	110	400	500	29	78	−47	−66
	220	450	510	24	79	−40	−60
	440	500	560	25	77	−23	−45
	950	570	640	23	70	−20	−13
SR	6	360	490	29	78	−61	−92
C₀	130	480	570	25	74	−45	−75
	230	510	580	23	75	−37	−68
	470	550	630	20	75	−19	−58
	910	600	680	20	69	7	−35

Note: AW – as-welded, SR – stress relieved at 580 °C.

$$YS_{sr} = 278 + 85\ Mn + 5626\ Nb - 35329\ Nb^2 \qquad [13.7]$$

$$TS_{sr} = 397 + 78\ Mn + 4316\ Nb - 25141\ Nb^2 \qquad [13.8]$$

Stress relief of the welds of higher niobium and manganese contents provided little or no softening, although no strengthening took place, as can occur with some niobium-containing steels on similar tempering. Titanium made little or no difference to the strength of as-welded deposits nor to the influence of niobium.

Charpy testing showed a markedly adverse effect of niobium on toughness of deposits containing ~40 ppm Ti in both as-welded and stress relieved conditions. Figure 13.18 shows increases in the 100 J transition temperature of around 60 °C as-welded and ~80 °C after stress relief with the highest niobium content. By and large, the optimum manganese content was maintained at ~1.4% for all niobium contents when titanium was present.

In the absence of added titanium (Table 13.6(c)), up to 200 ppm Nb improved the as-welded toughness of the 1.4%Mn deposits and

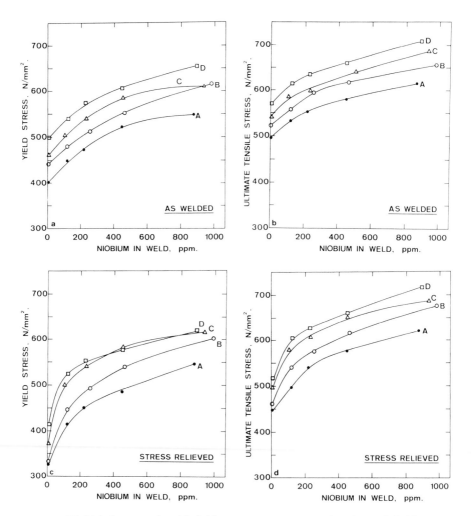

13.17 Influence of weld niobium content on strength values of C–Mn welds with 35 ppm Ti: *a*) Yield strength, as-welded; *b*) Tensile strength, as-welded; *c*) Yield strength, stress relieved at 580 °C; *d*) Tensile strength, stress relieved at 580 °C.

toughness at higher niobium levels was similar to the 1.4%Mn welds of the same niobium content with added titanium. Otherwise, as little as 200 (0.02%) ppm Nb gave an unacceptable increase in 100 J transition temperature after stress relief.

Despite the harmful effect of niobium, particularly after PWHT, the shift in transition temperature on stress relief of welds containing titanium was not uniform, as can be seen from Fig. 13.19. Niobium

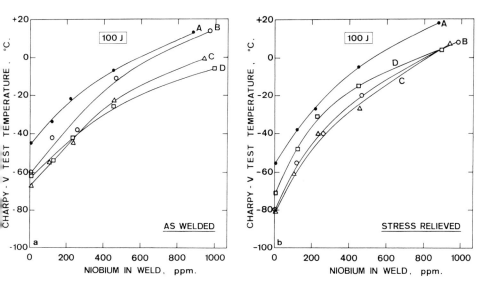

13.18 Influence of weld niobium content on Charpy transition temperature of deposits with varying manganese content and 35 ppm Ti: *a*) As-welded; *b*) After stress relief at 580 °C.

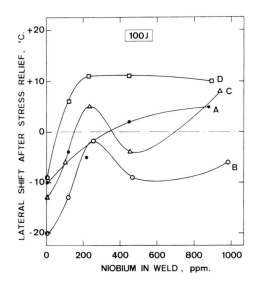

13.19 Influence of weld niobium content on shift in Charpy transition temperature on stress relief for deposits with varying manganese content and ~35 ppm Ti.

gave a positive shift in transition temperature in 1.8%Mn welds but a negative shift in 1%Mn.

Boron (B)

Boron is used to increase the hardenability of engineering steels, although its use in structural steels is limited to an obsolescent steel with 0.5%Mo. However, weld metals with titanium and boron have been used for a number of critical applications. Otherwise, normal weld metals contain no more than 1 or 2 ppm B. However, because of its low atomic weight (10.8, compared with 56 for iron), small amounts of boron in terms of normal weight percentages (or ppm) can have a large influence on microstructure and mechanical properties.

Table 13.7 Composition of weld metal with either 0.6 or 1.4%Mn having varying boron content

Element, wt%	C	Mn	Si	S	P	B	Ti	N	O
Electrode code									
A_0	0.063	0.66	0.36	0.007	0.007	0.0005	0.0005	0.006	0.048
	0.065	0.66	0.36	0.008	0.009	0.0030	0.0007	0.007	0.044
	0.063	0.65	0.39	0.008	0.009	0.0058	0.0003	0.007	0.045
	0.068	0.68	0.36	0.009	0.008	0.0099	0.0007	0.008	0.044
	0.064	0.66	0.43	0.010	0.010	0.013	0.0009	0.006	0.053
	0.066	0.69	0.40	0.008	0.008	0.017	0.0003	0.007	0.044
A	0.073	0.64	0.35	0.005	0.004	0.0003	0.0038	0.008	0.043
	0.066	0.67	0.38	0.008	0.010	0.0030	0.0034	0.007	0.054
	0.062	0.68	0.45	0.009	0.010	0.0067	0.0040	0.006	0.053
	0.061	0.67	0.45	0.009	0.010	0.0106	0.0042	0.006	0.052
	0.065	0.66	0.48	0.008	0.010	0.014	0.0044	0.006	0.051
	0.061	0.67	0.43	0.008	0.008	0.019	0.0045	0.006	0.048
C_0	0.073	1.51	0.30	0.007	0.007	0.0001	<0.0005	0.009	0.045
	0.067	1.53	0.27	0.008	0.007	0.0016	0.0002	0.009	0.046
	0.071	1.58	0.29	0.007	0.006	0.0028	0.0002	0.008	0.045
	0.066	1.48	0.33	0.008	0.007	0.0064	<0.0005	0.007	0.051
	0.066	1.48	0.37	0.008	0.008	0.0109	<0.0005	0.006	0.049
	0.068	1.53	0.39	0.008	0.007	0.015	<0.0005	0.008	0.049
	0.064	1.49	0.40	0.007	0.005	0.020	<0.0005	0.008	0.050
C	0.074	1.46	0.28	0.008	0.008	0.0003	0.0035	0.008	0.044
	0.068	1.51	0.34	0.007	0.008	0.0033	0.0033	0.007	0.047
	0.065	1.48	0.35	0.007	0.007	0.0070	0.0034	0.007	0.047
	0.064	1.43	0.36	0.007	0.007	0.011	0.0036	0.007	0.045
	0.067	1.45	0.31	0.008	0.008	0.014	0.0038	0.008	0.045
	0.066	1.44	0.39	0.007	0.009	0.019	0.0045	0.007	0.044

Note: all welds contained ≤5 ppm Al, Nb, V, ~0.03%Cr, Ni, Cu, ~0.005%Mo.

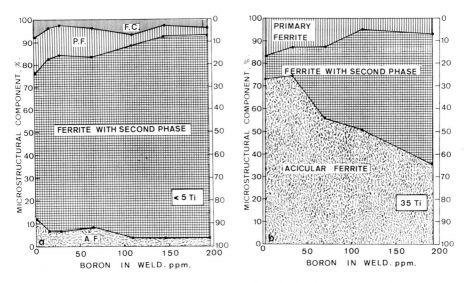

13.20 Influence of weld boron content on as-deposited microstructure of welds with 1.4%Mn and: *a*) <5 ppm Ti; *b*) 35 ppm Ti.

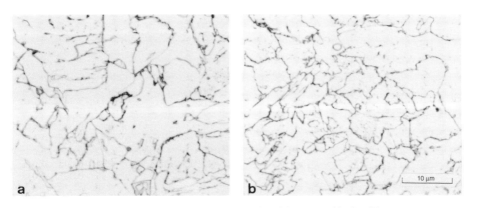

13.21 Microstructure of fine-grained regions of welds, as-welded, with 1.4%Mn and <5 ppm Ti with: *a*) 1 ppm B; *b*) 195 ppm B.

This section deals with the influence of boron itself on weld metals A_0, A, C_0 and C, containing nominal contents of 0.6 and 1.4%Mn at two levels of titanium, <5 and ~35 ppm (Ref. 3, Ch. 10), the boron being added via the coating as ferro-boron. Tensile tests were restricted to the as-welded condition, although Charpy impact tests were also carried out after PWHT. Weld metal compositions in Table 13.7 shows that all welds were well balanced.

Table 13.8 Mechanical properties of 1.4%Mn welds with varying boron content

Electrode Code	B, ppm	Yield strength, N/mm^2	Tensile strength, N/mm^2	Elongation, %	Reduction of area, %	100 J Charpy transition, °C	
						AW	SR
A$_0$	5	410	480	30	77	−53	−58
	30	400	480	30	70	−50	−60
	58	390	470	32	71	−53	−64
	99	370	450	30	75	−51	−67
	126	400	470	23	73	−57	−56
	170	380	480	26	74	−52	−58
A	3	400	490	33	78	−45	−49
	30	440	500	22	77	−35	−48
	67	440	480	21	71	−30	−43
	106	410	470	23	74	−42	−50
	140	410	470	24	76	−48	−59
	186	420	480	24	74	−44	−55
C$_0$	1	460	540	29	78	−15	−69
	16	460	530	26	78	−11	−68
	26	460	540	28	75	−14	−60
	64	460	540	22	76	−22	−50
	109	430	520	24	75	−37	−70
	146	470	550	21	73	−50	−69
	195	480	560	22	74	−34	−59
C	3	460	530	28	79	−68	−70
	33	510	580	25	76	−60	−74
	70	480	540	26	76	−55	−60
	112	480	530	25	75	−65	−73
	140	480	540	22	75	−60	−65
	192	460	560	23	74	−50	−61

Note: nd – not determined; AW – as-welded; SR – stress-relieved at 580 °C.

Metallography

Metallographic examination was confined to welds with 1.4%Mn. In the absence of titanium, boron increased the content of the major constituent, ferrite with aligned second phase, at the expense of all the other constituents – the first addition of 16 ppm B having the greatest effect (Fig. 13.20(*a*)). The titanium-containing weld metals (Fig. 13.20(*b*)), which in the absence of boron were of the acicular ferrite type, showed similar trends but much more markedly, so that the greatest boron addition of 200 ppm reduced the amount of acicular ferrite by half, to give a microstructure containing over 50% ferrite with second phase. In this case, however, the first small addition of 33 ppm B had virtually no effect on the proportions of the constituents.

These changes in as-deposited microstructure were reflected by those in coarse reheated regions. Adding boron altered the microstructure of fine-grained reheated regions from fine ferrite with aligned microphases to one with a precipitated sub-structure (Fig. 13.21).

Mechanical properties

The tensile test results in Table 13.8 show no obvious systematic effect of boron, either with or without a titanium addition to the weld, although welds with a small titanium addition were generally slightly stronger. This is illustrated by the plot of yield strength values in Fig. 13.22.

In the as-welded condition (Fig. 13.23(*a*)), boron additions up to 140 ppm improved the toughness of titanium-free 1.4%Mn deposits, reducing transition temperature by nearly 40 °C, although a further addition of boron to 200 ppm reduced toughness somewhat. However, in 1.4%Mn welds with 40 ppm Ti, boron additions up to 70 ppm had a slightly detrimental effect on toughness; 110 ppm B restored toughness almost to the boron-free level but further additions were again harmful. With 0.6%Mn and no titanium, boron had no effect on toughness, although with 40 ppm Ti, the behaviour mirrored that of the 1.4%Mn–Ti welds. As a result, the welds were less tough, and also less tough than the 0.6%Mn welds with no added titanium.

Stress relief heat treatment (Fig. 13.23(*b*)) greatly improved the

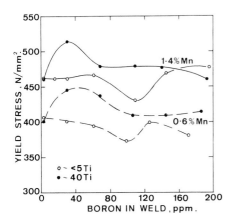

13.22 Influence of boron content on as-welded yield strength of 0.6 and 1.4%Mn welds with and without added titanium.

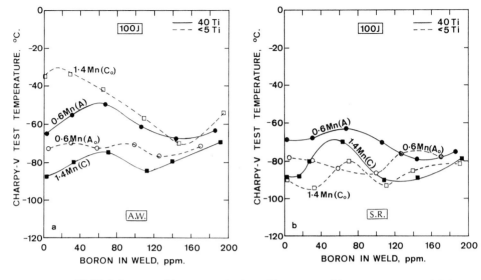

13.23 Influence of boron content on Charpy transition temperature of 0.6 and 1.4%Mn deposits with and without added titanium: *a*) As-welded; *b*) After stress relief at 580 °C.

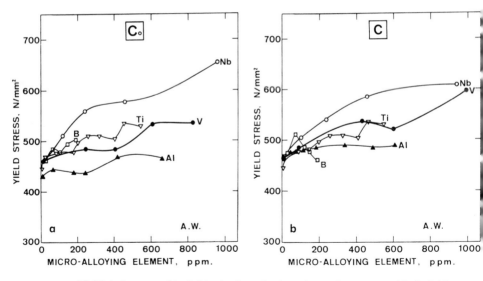

13.24 Influence of individual microalloying elements on as-welded yield strength of 1.4%Mn welds: *a*) Without added Titanium (C₀); *b*) With ~35 ppm Ti (C).

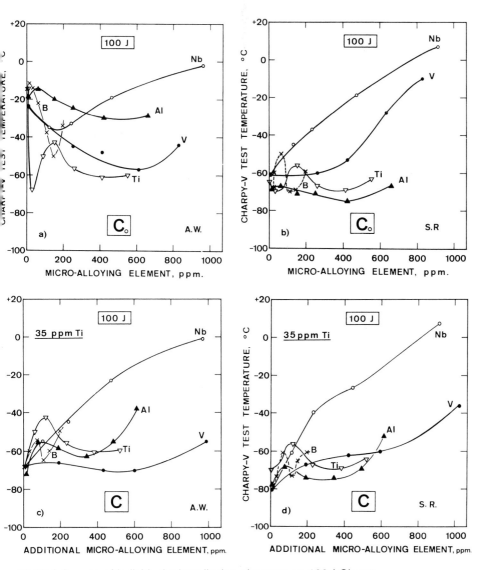

13.25 Influence of individual microalloying elements on 100 J Charpy transition temperature of 1.4%Mn welds: *a*) Without added Ti (C_0), as-welded; *b*) Without added Ti (C_0), after stress relief at 580 °C; *c*) With ~35 ppm Ti (C), as-welded; *d*) With ~35 ppm Ti (C), after stress relief at 580 °C.

toughness of titanium-free 1.4%Mn deposits to a level not markedly different from those containing titanium. The effects of boron on stress-relieved toughness were minor, although both 1.4%Mn series showed inferior toughness with 70 ppm B compared with any other boron content. This was mirrored to a lesser degree by the 0.6%Mn welds with 40 ppm Ti, although not with the Ti-free version.

General discussion

Effects of the individual elements studied in this chapter on as-welded yield strength of 1.4%Mn weld metals containing ~35 ppm Ti are compared with each other and that of titanium on Ti-free 1.4%Mn weld metal in Fig. 13.24. Although not all elements exerted a linear effect, they can be placed in order of increasing strengthening: Al, V, Ti and Nb. Boron was anomalous because, although strengthening in the absence of added titanium, its maximum addition did not strengthen deposits with 40 ppm Ti.

Effects of these elements on 100 J Charpy transition temperature are shown in Fig. 13.25 (Ref. 3, Ch. 10). Without titanium, as-welded toughness benefited from either the initial small additions or subsequent additions. The addition for optimum toughness varied from 35 ppm in the case of titanium to 500–600 ppm for vanadium and

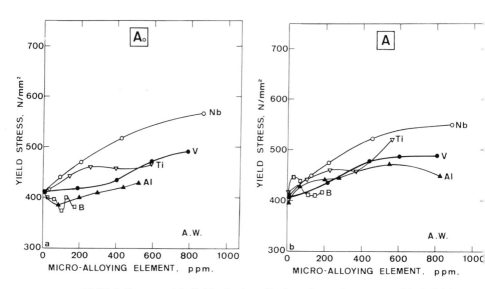

13.26 Influence of individual microalloying elements on as-welded yield strength of 0.6%Mn welds: *a*) Without added titanium (A₀); *b*) With ~35 ppm Ti (A).

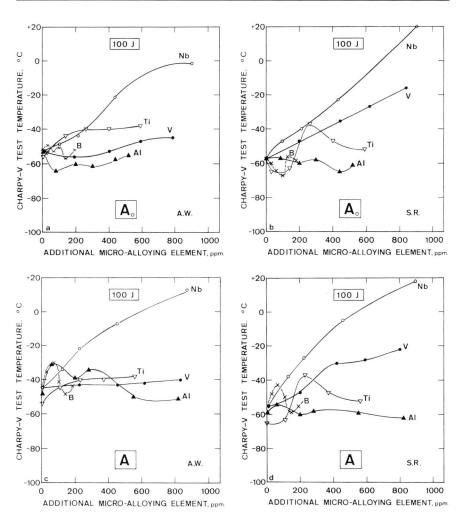

13.27 Influence of individual microalloying elements on 100 J Charpy transition temperature of 0.6%Mn welds: *a*) Without added Ti (A₀), as-welded; *b*) Without added Ti (A₀), after stress relief at 580 °C; *c*) With ~35 ppm Ti (A), as-welded; *d*) With ~35 ppm Ti (A), after stress relief at 580 °C.

aluminium. Titanium was exceptional in giving two optimum levels, the first at 35 ppm and the second at around 400 ppm.

In deposits already containing ~35 ppm Ti (Fig 13.25(*c*)), no additional element was wholly beneficial, although Al and V up to ~400 ppm and boron at about 150 ppm were not unhelpful. Niobium and (above 400 ppm) vanadium were definitely harmful and the other elements in excess appeared to be potentially damaging.

After stress relief at 580 °C (Fig. 13.25(b,d)) titanium had less influence on toughness. Aluminium, boron and titanium all initially increased transition temperature but then gave a reduction followed by another rise. Vanadium and, more particularly, niobium were always detrimental.

Although titanium alone was the most beneficial element with regard to toughness, combinations of aluminium, boron and titanium were considered to be worthy of further study and the results of these trials are examined in the next chapter.

The as-welded yield stresses of weld metals containing 0.6%Mn, without (A_0) and with (A) titanium are plotted in Fig. 13.26. Comparable trends were exhibited for the 1.4%Mn welds in Fig. 13.24, except in the case of boron, which was again anomalous.

Effects of the five microalloying elements on 100 J Charpy temperature of 0.6%Mn welds are shown in Fig. 13.27. Comparison of A_0 with C_0 in Fig. 13.25 shows a fundamental difference, with elements other than aluminium failing to induce a beneficial response in 0.6%Mn deposits at intermediate concentrations. Thus, it is clear that manganese is required at the higher level so that the microalloying elements can attain their specific effects. Apart from that, the trends shown in Fig. 13.25 and 13.27 are essentially comparable, with niobium, in particular, causing considerable embrittlement.

References

1 Evans GM: 'Influence of aluminium on the microstructure and properties of C–Mn all-weld-metal deposits'. *Weld J* 1991 **70**(1) 32s–39s; *Welding Res Abroad* 1992 **38**(8/9), 2–12; *Schweissmitteilungen* 1990 **48**(124) 15–31.
2 Evans GM: 'Effect of vanadium in manganese-containing MMA weld deposits'. *Welding Res Abroad* 1992 **38**(8/9) 23–32; *Schweissmitteilungen* 1991 **49**(126) 18–33.
3 Evans GM: 'Effect of niobium in manganese-containing MMA weld deposits'. *Welding Res Abroad* 1992 **38**(8/9) 33–44; *Schweissmitteilungen* 1991 **49**(127) 24–39.

14

Combinations of microalloying elements

Titanium and aluminium (Ti–Al)

Nucleation of acicular ferrite by some form of titanium compound associated with non-metallic inclusions is known from work on submerged-arc welds to be inhibited by excessive amounts of aluminium. This effect was discussed in the previous chapter at a low level ($\sim$35 ppm) of titanium. Although aluminium is rarely present in MMA welds, it was important to investigate these effects at higher titanium (and aluminium) levels, to obtain a better understanding of the mechanisms of acicular ferrite formation and inhibition.

Welds were made with 1.4%Mn (ideal for allowing acicular ferrite formation) and additions of up to 450 ppm Ti and 500 ppm Al and tested in the as-welded condition only.[1] Their compositions are detailed in Table 14.1; the composition and mechanical properties of welds with no titanium addition and with 35 ppm Ti were given in Tables 13.1 and 13.2 and 12.3 and 12.4, respectively.

Compositions were as well balanced as could be expected with two elements being added which were capable of reducing some of the components of the electrode coating. Silicon levels tended to increase as both titanium and aluminium were increased, whilst the nitrogen content in each series fell as aluminium was increased. Oxygen contents varied in an unexpected manner. Without aluminium, titanium reduced oxygen levels. This tendency was reduced and eventually nullified by aluminium, as shown in Fig. 14.1. With titanium increasing up to 80 ppm Ti, oxygen levels fell from $\sim$0.045 to $\sim$0.038% and were unaffected by aluminium. Greater amounts of titanium reduced oxygen still further in the absence of aluminium but as little as 50 ppm Al increased oxygen, so that with more than 100 ppm Al, oxygen levels were 0.04%, i.e. similar to those found with 40 ppm Ti or less.

Table 14.1 Composition of weld metal having varying titanium and aluminium content

Element, wt%	C	Mn	Si	S	P	Ti	Al	N	O
Ti, ppm									
80	0.074	1.50	0.32	0.006	0.006	0.010	0.0008	0.009	0.038
	0.074	1.41	0.27	0.007	0.006	0.009	0.0047	0.007	0.041
	0.071	1.36	0.39	0.005	0.009	0.009	0.016	0.005	0.038
	0.069	1.36	0.45	0.005	0.008	0.006	0.028	0.004	0.037
	0.069	1.39	0.51	0.004	0.007	0.007	0.048	0.004	0.038
	0.072	1.40	0.57	0.004	0.008	0.008	0.064	0.004	0.036
220	0.081	1.54	0.34	0.007	0.008	0.018	0.0003	0.008	0.031
	0.072	1.32	0.24	0.006	0.008	0.018	0.0047	0.007	0.040
	0.072	1.33	0.34	0.006	0.008	0.019	0.015	0.006	0.043
	0.074	1.36	0.41	0.005	0.005	0.026	0.032	0.005	0.042
	0.066	1.41	0.46	0.004	0.009	0.024	0.044	0.005	0.041
	0.066	1.33	0.44	0.004	0.008	0.024	0.068	0.005	0.045
450	0.069	1.49	0.43	0.004	0.005	0.046	0.0001	0.008	0.029
	0.071	1.44	0.36	0.007	0.005	0.046	0.0067	0.009	0.037
	0.071	1.39	0.42	0.007	0.005	0.042	0.015	0.006	0.045
	0.071	1.44	0.52	0.005	0.005	0.046	0.030	0.007	0.047
	0.069	1.41	0.54	0.005	0.005	0.046	0.042	0.006	0.044
	0.067	1.40	0.57	0.005	0.005	0.045	0.050	0.006	0.045

Notes: all welds contained <5 ppm B, Nb, V, ~0.005%Mo, ~0.03%Cr, Ni, Cu.
For compositions of welds with no added Ti and 35 ppm Ti see Tables 13.1 and

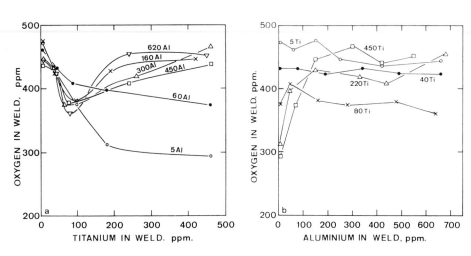

14.1 Influence on the oxygen content of 1.4%Mn weld metals of: *a)* Titanium; *b)* Aluminium content.

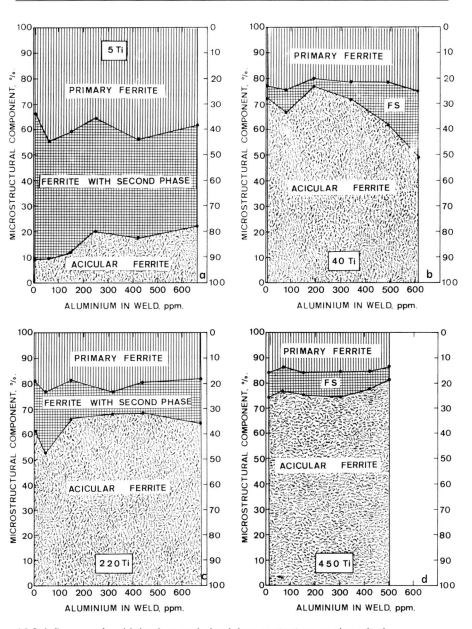

14.2 Influence of weld titanium and aluminium content on as-deposited microstructure of welds with 1.4%Mn.

Metallography

The as-deposited microstructure of the selected welds exhibits little change with aluminium content (Fig. 14.2), except for the series with 40 ppm Ti, which was discussed in the section on aluminium in the previous chapter. The overall effect of titanium was to increase the proportion of acicular ferrite at the expense of the other two constituents. However, as described earlier for aluminium-free weld metals in Fig. 13.1, intermediate titanium contents around 100 ppm showed less acicular ferrite than those of lower or higher titanium level; this trend was apparent in Fig. 14.2 at most aluminium levels. With 220 and 450 ppm Ti, aluminium had little effect on microstructure and the high proportion of acicular ferrite in deposits of highest

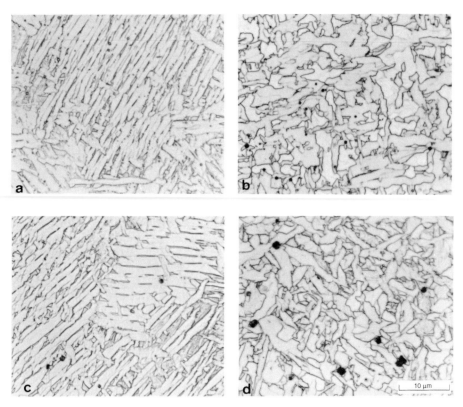

14.3 Microstructure of as-deposited regions of welds with: *a*) 5 ppm Ti and 6 ppm Al; *b*) 460 ppm Ti and 1 ppm Al; *c*) 4 ppm Ti and 660 ppm Al; *d*) 450 ppm Ti and 500 ppm Al.

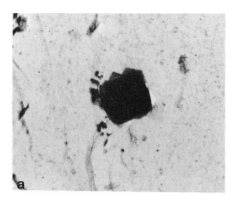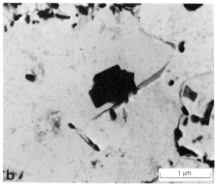

14.4 TEM micrographs of: *a*) Cubic; *b*) Multiple particles, in reheated regions of 1.4%Mn welds with 450 ppm Ti and 500 ppm Al.

aluminium content is noteworthy, as the aluminium was in excess of that required to form Al_2O_3.

The microstructure (Fig. 14.3) of weld metal without titanium was essentially bainitic, whether or not aluminium was present, whilst those with sufficient titanium had characteristic acicular ferrite microstructures – even in the presence of high aluminium contents. The deposit with high contents of both aluminium and titanium (Fig. 14.3(*d*)) also contained high proportions of both microphases and cubic particles.

The coarse-grained reheated regions contained constituents similar to the as-deposited regions. In addition to refining the fine-grained regions, the two microalloying elements introduced changes already discussed. In the highest alloyed weld, cubic and multiple particles (Fig. 14.4) were seen; these are discussed in Chapter 17.

Mechanical properties

The mechanical properties of deposits with no added titanium and 36 ppm Ti were given in Tables 13.2 and 12.4; those of deposits with higher titanium content are in Table 14.2. The tensile test results (Fig. 14.5) show a general trend for the yield strength to increase steadily as weld titanium content increased, with only a slight influence of aluminium. It is likely, however, that some of the increase is a result of the increase in silicon content (~0.1 to ~0.2%) as aluminium and titanium were increased. One undoubted feature of Fig. 14.5 is that the first small titanium addition gave a sharp increase in strength and,

Table 14.2 Mechanical properties, as-welded, of 1.4%Mn welds with varying titanium and aluminium content

Ti level	Al, ppm	Yield strength, N/mm²	Tensile strength, N/mm²	Elongation, %	Reduction of area, %	Charpy transition, °C	
						100 J	28 J
80	8	480	550	29	80	−52	−74
	50	440	510	32	82	−31	−57
	160	480	540	31	81	−56	−68
	280	470	540	28	81	−61	−68
	480	470	530	27	77	−54	−70
	640	490	560	27	78	−41	−68
220	3	500	560	27	79	−52	−67
	50	450	540	31	79	−35	−54
	150	470	520	29	78	−52	−64
	320	510	570	24	77	−47	−63
	440	520	580	24	76	−42	−68
	680	520	590	24	72	−42	−76
450	1	540	590	26	80	−60	−78
	70	520	580	22	79	−48	−69
	150	530	590	25	78	−56	−74
	230	540	600	23	77	−49	−65
	420	560	620	26	77	−43	−63
	500	550	620	23	75	−44	−76

Note: for properties of welds with no added Ti and 36 ppm Ti see Tables 13.2 and 12.4.

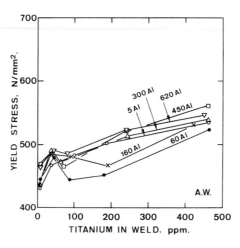

14.5 Influence of weld titanium and aluminium content on as-welded yield strength of 1.4%Mn deposits.

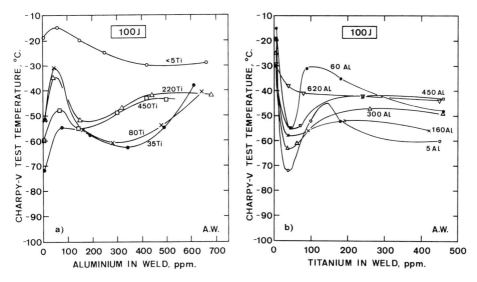

14.6 Influence on 100 J Charpy transition temperature of as-welded deposits with 1.4%Mn of: *a)* Aluminium; *b)* Titanium.

in all except the aluminium-free welds, the small peak in strength was followed by a slight fall. Despite the trend of silicon to increase with aluminium, the aluminium-free welds were not the weakest and the welds of highest aluminium content were not always the strongest.

Charpy impact test results are presented in Fig. 14.6 to show the effects of titanium and aluminium on the 100 J temperature. In every case, addition of a small amount – ~50 ppm – of aluminium impaired toughness to a greater or smaller degree, the worst embrittlement being with intermediate titanium levels of 80 and 220 ppm. Further additions of aluminium improved toughness but, with all except the titanium-free deposits, a further deterioration then took place.

With titanium additions to welds of constant aluminium content, the situation was reversed; the smallest addition of ~40 ppm gave a sharp improvement, subsequent increases worsening toughness and then, in several cases, giving a further improvement. An exception to this was the deposits of the highest aluminium level (620 ppm), where titanium gives a smaller improvement with the first addition and this improvement continued to the maximum titanium level of 450 ppm.

A three-dimensional model of the behaviour of 100 J temperature in Fig. 14.7 shows a valley of good toughness rising from the Ti-axis at ~40 ppm Ti to peter out at the 620 ppm Al level, with a side valley

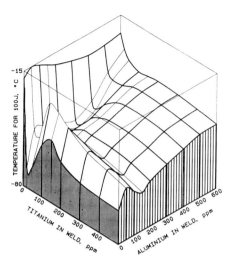

14.7 Three-dimensional model showing joint influences of aluminium and titanium on 100 J Charpy transition temperature of as-welded deposits with 1.4%Mn.

originating at ~40 ppm Ti-350 ppm Al, rising and then falling with increasing titanium and decreasing aluminium content. The harmful influence of ~50 ppm Al can be seen in the ridge running parallel to the titanium axis.

Discussion

The complexity of the behaviour, particularly toughness, of weld metals containing both titanium and aluminium requires care both in selection of raw materials for welding consumables for high quality applications and in devising procedures for welding steels containing aluminium in cases where high dilution may occur. The situation is further complicated if consideration is given to the influence of these elements at different levels of other elements such as manganese, boron and nitrogen, some of which are considered later in this chapter.

Titanium and boron (Ti–B)

The combined effects of titanium and boron have been used in submerged-arc and flux-cored arc (FCAW) welding to improve toughness[2-7] and the present work[8] studied their joint effects on weld

Table 14.3 Composition of 1.4%Mn weld metal having varying titanium and boron content

Element, wt%	C	Mn	Si	S	P	B	Ti	N	O
Ti, ppm									
120	0.074	1.45	0.30	0.007	0.009	0.0002	0.012	0.008	0.034
	0.071	1.55	0.35	0.007	0.010	0.0038	0.013	0.007	0.037
	0.071	1.53	0.34	0.007	0.010	0.0071	0.011	0.007	0.038
	0.068	1.45	0.34	0.008	0.009	0.0112	0.013	0.008	0.034
	0.066	1.43	0.33	0.008	0.009	0.0148	0.012	0.008	0.035
	0.067	1.41	0.36	0.009	0.009	0.0195	0.012	0.008	0.037
260	0.072	1.50	0.38	0.006	0.010	0.0002	0.026	0.008	0.029
	0.070	1.48	0.43	0.007	0.010	0.0021	0.026	0.007	0.031
	0.070	1.53	0.42	0.007	0.009	0.0046	0.029	0.007	0.029
	0.066	1.45	0.39	0.006	0.008	0.0091	0.027	0.008	0.026
	0.074	1.42	0.38	0.006	0.009	0.0126	0.027	0.007	0.027
	0.069	1.36	0.37	0.007	0.009	0.0169	0.025	0.006	0.025
400	0.069	1.47	0.45	0.005	0.006	0.0002	0.041	0.008	0.028
	0.077	1.56	0.39	0.006	0.011	0.0014	0.040	0.007	0.029
	0.074	1.50	0.39	0.006	0.009	0.0029	0.041	0.007	0.028
	0.075	1.49	0.37	0.006	0.010	0.0041	0.037	0.007	0.028
	0.075	1.50	0.40	0.006	0.010	0.0051	0.039	0.008	0.030
	0.077	1.53	0.42	0.006	0.010	0.0108	0.043	0.008	0.030
	0.073	1.52	0.40	0.006	0.011	0.0158	0.039	0.008	0.029
600	0.074	1.50	0.47	0.006	0.010	0.0001	0.058	0.008	0.029
	0.079	1.54	0.48	0.006	0.010	0.0014	0.061	0.008	0.028
	0.073	1.54	0.48	0.006	0.010	0.0032	0.052	0.008	0.029
	0.075	1.52	0.45	0.005	0.009	0.0059	0.061	0.008	0.028
	0.070	1.49	0.42	0.006	0.009	0.0095	0.061	0.006	0.030
	0.067	1.48	0.40	0.005	0.009	0.0122	0.064	0.008	0.030
	0.074	1.49	0.39	0.005	0.009	0.0158	0.066	0.007	0.032

Notes: all welds contained <5 ppm Al, Nb, V, ~0.005%Mo, ~0.03%Cr, Ni, Cu.
For composition of welds with 5 and 35 ppm Ti see Table 13.7.

metal strength in the as-welded condition and Charpy impact toughness as-welded and after stress relief at 580 °C. The base composition was 1.4%Mn with additions of up to 600 ppm Ti and up to 200 ppm B. Tests on welds of different boron content with 5 and 35 ppm Ti have already been described in the section on boron in the previous chapter.

The composition of welds additional to those with 5 and 35 ppm Ti (which were given in Table 13.7) is detailed in Table 14.3. Most elements were kept constant except for silicon, which tended to increase with increasing titanium content, and oxygen, which fell as the titanium increased and then stabilised at ~300 ppm with ≥300 ppm Ti. This behaviour is shown in Fig. 14.8, in which the width

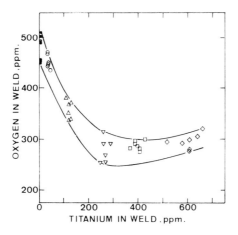

14.8 Influence of weld titanium content on weld oxygen level of 1.4%Mn deposits with up to 200 ppm B.

of the scatter band was due to the small increases in oxygen content as boron was increased, as noted in Chapter 13.

Metallography

The results of point counting for the topmost (un-reheated) runs are shown in Fig. 14.9; those for the two lowest titanium content welds were given in Fig. 13.20. The tendency of boron to increase the proportion of ferrite with second phase was continued with increasing titanium content but with an important exception, which was beginning to be seen in the 35 ppm Ti weld in Fig. 13.20(b). With 120 ppm Ti and more, the first additions of boron *increased* the amount of acicular ferrite at the expense of ferrite with aligned second phase and primary ferrite, so that in deposits containing 260 and 400 ppm Ti and ~30–40 ppm B, the acicular ferrite content of over 80% exceeded that of the weld with 35 ppm Ti and no boron. Only with higher boron contents at these titanium levels did ferrite with second phase increase at the expense of both acicular and primary ferrite.

The overall three-dimensional view of as-deposited microstructure in Fig. 14.10 reveals clearly how the high acicular ferrite content of the C–Mn–Ti weld metals develops into a ridge as 30–40 ppm B is added and then slumps as boron is increased. However, relatively high acicular ferrite contents persisted to high boron levels with titanium contents between 100 and 300 ppm. Figure 14.10 also shows that the smallest addition of titanium caused ferrite-carbide aggregate to dis-

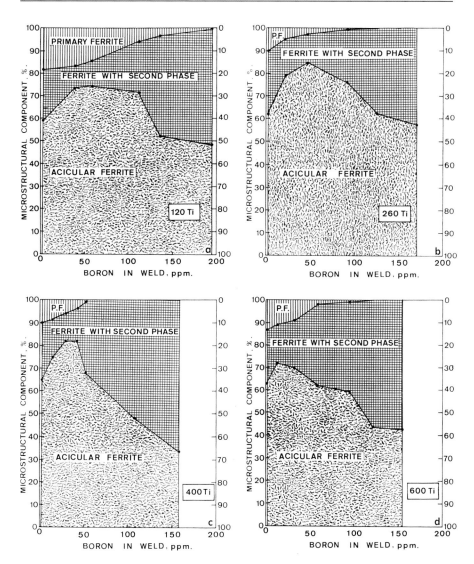

14.9 Influence of weld boron content on as-deposited microstructure of welds with 1.4%Mn and: *a*) 120 ppm Ti; *b*) 260 ppm Ti; *c*) 400 ppm Ti; *d*) 600 ppm Ti; see Fig. 13.19 for deposits with 5 and 35 ppm Ti.

appear over the whole range of boron contents examined. Primary ferrite decreased as boron and titanium were increased and was absent from most weld metals containing >100 ppm B and >300 ppm Ti.

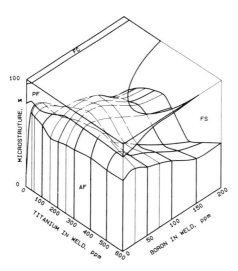

14.10 Three-dimensional model showing joint influences of boron and titanium on as-deposited microstructure of welds with 1.4%Mn. (AF – acicular ferrite, FC – ferrite-carbide aggregate, FS – ferrite with second phase, PF – primary ferrite).

The transformed austenite grain centres of the coarse-grained reheated weld metal again had structures resembling those of as-deposited weld metal. The refined structure of fine-grained reheated regions contained more M/A constituent as amounts of boron and titanium were increased.

Mechanical properties

A very complex situation is apparent when strength values at different titanium levels (Table 14.4) are plotted against boron content (Fig. 14.11). Titanium generally appeared to have strengthened the deposits, although this may have been partly a result of silicon increasing with titanium. Boron had a complex effect; in the presence of titanium, small additions (~20 ppm) gave a small increase in strength (apparent even in the 40 ppm Ti weld in Fig. 13.22). Except for the deposit of highest titanium content, larger additions of boron gave a weakening followed by a strengthening. A minimum in strength occurred at lower boron contents as the titanium content increased, so that the 600 ppm Ti weld exhibited only a slight step on the curve of increasing yield strength.

Charpy tests carried out both as-welded and after PWHT are sum-

Table 14.4 Mechanical properties of 1.4%Mn welds with varying boron and titanium content

Ti ppm	B ppm	Yield strength, N/mm²	Tensile strength, N/mm²	Elongation, %	Reduction of area, %	100 J Charpy transition, °C	
						AW	SR
120	2	480	540	28	80	−43	−56
	38	510	560	26	78	−57	−58
	70	490	540	28	74	−56	−60
	110	490	550	25	74	−65	−63
	150	490	570	24	72	−59	−60
	200	480	550	25	74	−40	−40
260	2	510	580	26	79	−56	−67
	21	530	580	26	78	−72	−72
	46	500	560	26	75	−82	−78
	90	500	560	28	76	−81	−66
	130	490	570	27	78	−69	−59
	170	480	560	26	74	−52	−50
400	2	500	580	26	80	−61	−62
	14	560	620	24	74	−77	−71
	29	550	590	25	75	−79	−75
	41	550	590	26	73	−81	−81
	51	510	570	28	79	−83	−76
	110	550	600	25	74	−62	−69
	160	520	590	22	78	−53	−44
600	1	550	620	23	76	−56	−60
	14	540	620	25	77	−74	−66
	32	520	590	26	75	−76	−78
	60	540	610	21	71	−65	−69
	100	560	610	22	74	−55	−57
	120	550	600	24	73	−42	−47
	160	540	600	24	73	−26	−22

Notes: for properties of welds with 5 and 35 ppm Ti see Table 13.8.
AW – as-welded; SR – stress-relieved.

marised in Fig. 14.12. As-welded, the deposits with significant titanium contents (i.e. above 35 ppm which, together with the titanium-free weld, were discussed in the section on boron in Chapter 13) showed an improvement in toughness as boron was added, followed by embrittlement. The minimum transition temperatures (like the minimum strengths in Fig. 14.11) occurred at lower boron contents as the titanium content increased. The best toughness values (100 J transition temperatures below −80 °C) were obtained around 40–50 ppm B with 260–400 ppm Ti – the compositions having the

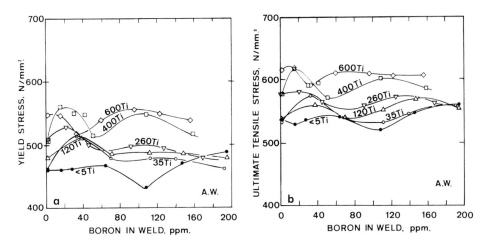

14.11 Influence of weld boron content on as-welded deposits with 1.4%Mn at varying titanium levels: *a*) Yield strength; *b*) Tensile strength.

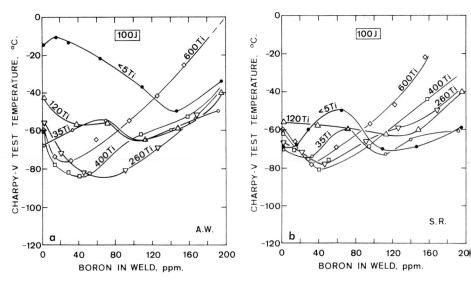

14.12 Influence of weld boron content on 100 J Charpy transition temperature of 1.4%Mn deposits with varying titanium levels: *a*) As-welded; *b*) Stress-relieved at 580 °C.

maximum proportion of acicular ferrite in the as-deposited weld metal (Fig. 14.9 and 14.10).

A three-dimensional toughness model of as-welded transition temperature (Fig. 14.13(*a*)), reveals a valley of good toughness rising

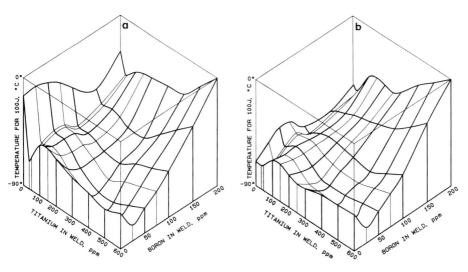

14.13 Three-dimensional models showing joint influences of boron and titanium on 100 J Charpy transition temperature of welds with 1.4%Mn: *a*) As-welded; *b*) Stress-relieved at 580 °C.

from the titanium-axis at about 40 ppm Ti with a side channel at right angles running down to the boron-axis at about 25 ppm B. Perpendicular to this side valley (at about 400 ppm Ti) a further shallow dip, hardly discernible on the Ti-axis, runs upwards to the back wall.

Stress relief (Fig. 14.12(*b*)) improved the toughness of the least tough welds (particularly the titanium-free and the 600 ppm Ti deposits with high boron levels) without changing the general pattern, which can be seen in the three-dimensional model (Fig. 14.13(*b*)). Again, the best toughness (100 J Charpy temperature at about − 80 °C) occurred with 260–400 ppm Ti and about 40 ppm B.

Titanium, boron and aluminium (Ti–B–Al)

Effects of aluminium on C–Mn–Ti–B weld metals were studied with a 1.4%Mn deposit having a relatively high titanium level of 450 ppm with 50 ppm boron,[9] i.e. for welds exhibiting optimum toughness in the previous section. Tests were restricted to the as-welded condition for tensile properties but Charpy tests were also carried out after stress relief at 580 °C. The compositions of the welds in Table 14.5 show that silicon increased with increasing aluminium whilst nitrogen tended to be reduced, possibly as a result of aluminium vapour in the arc atmosphere reducing the partial pressure of nitrogen.

Table 14.5 Composition of 1.4%Mn weld metal with titanium and boron having varying aluminium content

Element, wt%

C	Mn	Si	S	P	Ti	B	Al	N	O
0.070	1.57	0.45	0.006	0.010	0.039	0.0039	0.0013	0.083	0.031
0.074	1.53	0.44	0.007	0.010	0.045	0.0048	0.0068	0.082	0.037
0.072	1.56	0.49	0.007	0.010	0.042	0.0048	0.016	0.067	0.044
0.071	1.51	0.57	0.007	0.012	0.046	0.0051	0.031	0.060	0.048
0.072	1.53	0.60	0.006	0.010	0.044	0.0054	0.038	0.053	0.047
0.078	1.44	0.60	0.006	0.007	0.054	0.0056	0.058	0.041	0.044

Note: all welds contained ~0.03%Cr, Ni and Cu, 0.005%Mo, <0.0005%Nb and V. For compositions of welds with <5 ppm B, see Table 12.3.

Oxygen, which was initially low because of the high titanium level, increased as aluminium was added, as would be expected from Fig. 14.1, roughly to the level expected from weld metal of low titanium content.

Microstructure

Quantitative metallography of the as-deposited microstructure (Fig. 14.14) revealed that more than 160 ppm Al increased the proportion of ferrite with second phase at the expense of acicular and primary ferrite, the increase steepening above 380 ppm Al. This behaviour was in contrast to the effect of aluminium in a similar high-titanium weld metal without boron (Fig. 14.2(*d*)), where more than 300 ppm Al reduced the proportions of both ferrite with second phase and primary ferrite. The behaviour is more akin to the effect of aluminium on a deposit with 40 ppm Ti (Fig. 14.2(*b*)), except that the boron addition has reduced the amount of primary ferrite almost to zero with the maximum aluminium content.

Aluminium up to 160 ppm refined the acicular ferrite, whilst further additions coarsened it (Fig. 14.15 and 14.16). Higher magnification reveals the presence of cubic particles, shown in Chapter 17 to be Al–Ti compounds. The coarse-grained reheated microstructure resembled the as-deposited regions in that aluminium initially refined and then coarsened the microstructure and, at the same time, reduced the proportion of primary ferrite. Both fine- and coarse-grained reheated regions developed aligned microphases as aluminium was increased.

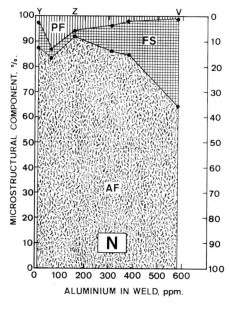

14.14 Influence of aluminium on as-deposited microstructure of welds with 1.4%Mn, 400 ppm Ti and 40 ppm B.

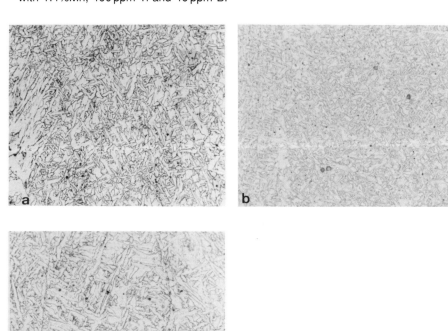

14.15 Microstructure of as-deposited regions of 1.4%Mn welds with 400 ppm Ti, 40 ppm B and: *a*) 13 ppm Al; *b*) 160 ppm Al; *c*) 580 ppm Al.

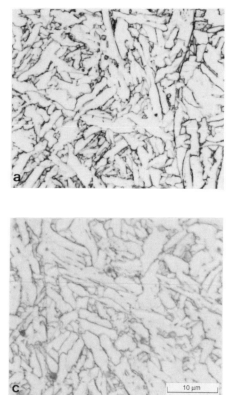

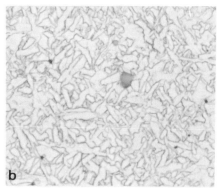

14.16 Microstructure of as-deposited regions of 1.4%Mn welds with 400 ppm Ti, 40 ppm B and: *a*) 13 ppm Al; *b*) 160 ppm Al; *c*) 580 ppm Al.

Mechanical properties

From the summary of mechanical properties in Table 14.6, it is apparent that up to 400 ppm Al strengthened the weld metal by over 100 N/mm^2; increasing aluminium to 600 ppm has no further effect. It can be seen from Fig. 14.17 that the presence of 50 ppm B enabled the bulk of this strengthening to occur; in the absence of boron, aluminium raised the as-welded strength by little more than 20 N/mm^2.

The influence of aluminium on toughness was also enhanced by boron (Fig. 14.18). As-welded with 40 ppm B, up to 160 ppm Al had little effect and the 100 J Charpy temperature remained below –80 °C. Subsequent additions up to 600 ppm Al brought the transition temperature up to nearly –10 °C, although without boron, similar addition of aluminium changed the transition temperature by less than 20 °C. In the stress-relieved condition, even the first addition of

Table 14.6 Mechanical properties of 1.4%Mn welds with 40 ppm Ti, 50 ppm B and varying aluminium contents

Al, ppm	Yield strength, N/mm²	Tensile strength, N/mm²	Elongation, %	Reduction of area, %	100 J Charpy transition, °C	
					AW	SR
13	550	590	26	73	−84	−81
68	580	600	27	76	−82	−68
160	610	640	27	73	−83	−61
310	600	660	23	74	−48	−41
380	680	740	21	70	−37	−46
580	670	730	20	70	−12	−2

Notes: for properties of welds with <5 ppm B see Table 12.4.
AW – as-welded; SR – stress-relieved.

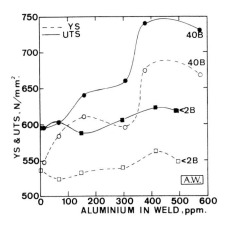

14.17 Influence of aluminium content on strength of as-welded deposits with 1.4%Mn and 400 ppm Ti with and without 40 ppm B.

70 ppm Al had a detectable effect and the 100 J temperature was almost 0 °C by the time 600 ppm Al had been added.

Titanium, boron and nitrogen (Ti–B–N)

To examine the influence of increasing nitrogen content on 1.4%Mn weld metals with titanium and boron additions, the welds whose compositions were given in Tables 13.7 and 14.3 were taken as a basis

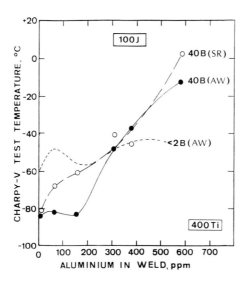

14.18 Influence of aluminium content on 100 J Charpy transition temperature of deposits with 1.4%Mn and 400 ppm Ti, with and without 40 ppm B.

and additions of nitrided manganese were made to the electrode coatings to achieve levels of ~160 and 240 ppm N in the welds (coded N1 and N2, respectively) (Ref. 3, Ch. 12). Compositional details of the additional welds are given in Table 14.7. Increasing nitrogen had no effect on content of titanium or boron, although it did reduce the level of oxygen at lower titanium content, thus reducing slightly the difference in oxygen content between the low and high titanium levels, which was found earlier in Tables 13.7 and 14.3.

The matrix of experimental compositions is shown in Fig. 14.19; particular attention was paid to welds of the following composition:

O   no additions
W   35 ppm Ti
X   400 ppm Ti
Y   400 ppm Ti: 40 ppm B
T   200 ppm B
U   400 ppm Ti: 160 ppm B.

Metallography

The proportions of different constituents in the top, unreheated beads of the deposits of highest and lowest nitrogen contents are given in

Table 14.7 Compositions of 1.4%Mn weld metal having varying titanium, boron and nitrogen content

Element, wt%

N	Ti, ppm	C	Mn	Si	S	P	Ti	B	N	O
160	120	0.070	1.50	0.29	0.008	0.009	0.0120	0.0002	0.016	0.034
		0.071	1.53	0.27	0.005	0.009	0.0120	0.0032	0.017	0.032
		0.073	1.53	0.28	0.005	0.010	0.0130	0.0057	0.017	0.031
		0.068	1.47	0.27	0.005	0.009	0.0110	0.0090	0.015	0.027
		0.068	1.43	0.29	0.005	0.009	0.0110	0.0129	0.015	0.031
		0.068	1.45	0.35	0.005	0.009	0.0100	0.0158	0.017	0.031
160	250	0.068	1.51	0.39	0.007	0.009	0.0300	0.0002	0.017	0.028
		0.071	1.54	0.37	0.004	0.007	0.0200	0.0046	0.016	0.029
		0.070	1.51	0.42	0.005	0.008	0.0210	0.0105	0.016	0.027
		0.070	1.61	0.39	0.004	0.006	0.0210	0.0147	0.017	0.027
		0.067	1.50	0.41	0.004	0.006	0.0230	0.0200	0.014	0.027
160	400	0.066	1.48	0.47	0.007	0.011	0.0410	0.0002	0.016	0.029
		0.072	1.53	0.37	0.006	0.011	0.0370	0.0015	0.015	0.029
		0.071	1.51	0.38	0.007	0.011	0.0410	0.0029	0.014	0.029
		0.069	1.48	0.34	0.007	0.010	0.0370	0.0040	0.015	0.028
		0.068	1.46	0.34	0.006	0.010	0.0350	0.0054	0.014	0.028
		0.069	1.48	0.35	0.007	0.011	0.0350	0.0102	0.014	0.028
		0.066	1.39	0.35	0.006	0.009	0.0360	0.0158	0.014	0.029
240	120	0.070	1.45	0.28	0.009	0.009	0.0120	0.0002	0.024	0.032
		0.068	1.46	0.35	0.005	0.010	0.0120	0.0029	0.024	0.031
		0.068	1.45	0.34	0.005	0.009	0.0120	0.0059	0.025	0.034
		0.068	1.39	0.33	0.005	0.009	0.0110	0.0090	0.027	0.031
		0.065	1.47	0.30	0.004	0.009	0.0110	0.0128	0.024	0.032
		0.061	1.37	0.30	0.005	0.009	0.0100	0.0162	0.021	0.031
240	250	0.067	1.48	0.40	0.006	0.008	0.0320	0.0002	0.025	0.029
		0.072	1.47	0.32	0.005	0.006	0.0190	0.0043	0.027	0.025
		0.069	1.41	0.34	0.005	0.006	0.0210	0.0091	0.024	0.026
		0.068	1.44	0.34	0.005	0.006	0.0210	0.0143	0.024	0.027
		0.070	1.37	0.40	0.006	0.006	0.0250	0.0200	0.022	0.027
240	400	0.070	1.46	0.40	0.007	0.010	0.0430	0.0002	0.023	0.030
		0.071	1.51	0.38	0.007	0.010	0.0410	0.0016	0.023	0.029
		0.068	1.42	0.36	0.007	0.009	0.0380	0.0028	0.022	0.030
		0.069	1.51	0.36	0.007	0.008	0.0410	0.0044	0.023	0.029
		0.071	1.46	0.36	0.007	0.010	0.0370	0.0052	0.023	0.028
		0.073	1.44	0.35	0.007	0.011	0.0360	0.0110	0.023	0.029
		0.066	1.40	0.36	0.007	0.012	0.0390	0.0167	0.022	0.030

Notes: all welds contained ~0.03%Cr, Ni and Cu, 0.005%Mo, <5 ppm Al, Nb and V. For compositions of welds with 80 ppm N see Tables 13.7 and 14.3.

Fig. 14.20 in the form of three-dimensional models. For clarity, these show separately the effect of Ti and B at two nitrogen levels (80 and 240 ppm) on ferrite with second phase, FS, in Fig. 14.20 (*a*) and (*c*) and other constituents in Fig. 14.20 (*b*) and (*d*). Boron and nitrogen together reduced the sharp fall in the proportion of FS on adding titanium, so that with high boron and nitrogen, the fall continued steadily to over 250 ppm Ti. High boron and nitrogen also removed

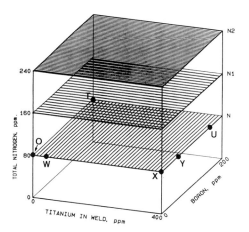

14.19 Matrix of experimental 1.4%Mn compositions with varying Ti, B and N, showing coded welds selected for special consideration.

the hump (accentuated by nitrogen) in FS at around 120 ppm Ti, whilst high nitrogen also removed the valley of low FS at about 40 ppm B.

The small amount of ferrite carbide aggregate (FC), which persisted to the highest boron level when nitrogen was low, was completely absent from any of the higher nitrogen deposits. Primary ferrite (PF) persisted in small amounts over the whole range of compositions with high nitrogen, although it was absent from welds of normal nitrogen content when both boron and titanium were high. Nitrogen also reduced the amount of increase in FS as boron was increased, so that the proportion of acicular ferrite (AF) was actually higher in welds with high boron, titanium and nitrogen content than when nitrogen was low.

The photomicrographs in Fig. 14.21 show some of the changes described above. Acicular ferrite in weld X, with high titanium and no boron, was increased by increasing the nitrogen content (Fig. 14.21 (*a*) and (*b*)). The same composition (weld Y) with 40 ppm B showed an increase in FS and primary ferrite when nitrogen was increased. With the highest boron level (weld U); increasing nitrogen gave less FS, more acicular ferrite and some ferrite veining.

These changes were parallelled in the coarse-grained reheated regions, with the additional feature that high nitrogen contents suppressed the grain boundary decoration in the deposits containing boron and gave better defined ferrite envelopes to the prior austenite grains, as shown in Fig. 14.22.

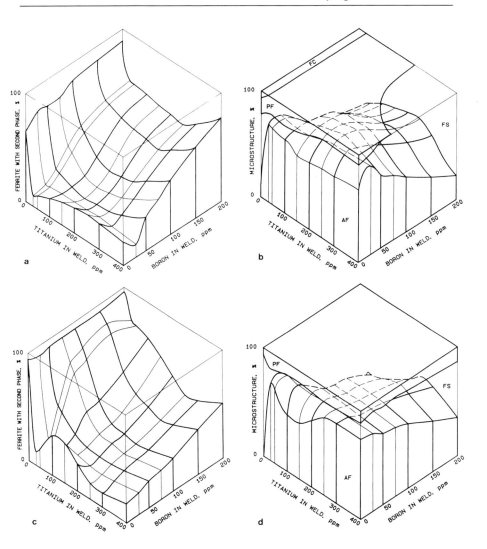

14.20 Influence of weld boron and titanium content on as-deposited microstructure of welds with 1.4%Mn and: *a,b*) Normal nitrogen level; *c,d*) 250 ppm N; *a,c*) Show proportions of ferrite with aligned second phase; *b,d*) Show acicular ferrite and other constituents.

The grain size of the fine-grained reheated regions was not greatly influenced by nitrogen. The quantity of micro-phases was increased by nitrogen when boron was absent (Fig. 14.23 (*a*) and (*b*)), whilst increasing nitrogen in deposits of low boron content induced

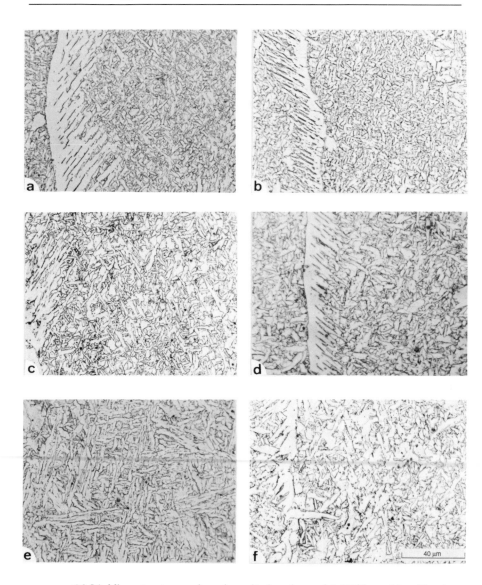

14.21 Microstructures of as-deposited regions of 1.4%Mn welds with: *a*) 400 ppm Ti and 80 ppm N (X); *b*) 450 ppm Ti and 250 ppm N (X2); *c*) 400 ppm Ti, 40 ppm B and 80 ppm N (Y); *d*) 400 ppm Ti, 40 ppm B and 230 ppm N (Y2); *e*) 400 ppm Ti, 160 ppm B and 80 ppm N (U); *f*) 400 ppm Ti, 170 ppm B and 220 ppm N (U2).

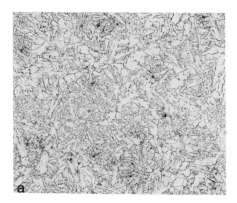
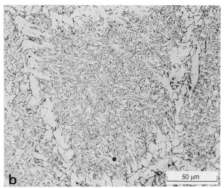

14.22 Microstructure of coarse-grained reheated regions of welds with: *a*) 400 ppm Ti, 40 ppm B and 80 ppm N (Y); *b*) 400 ppm Ti, 40 ppm B and 230 ppm N (Y2).

formation of cementite at grain boundaries, as shown in Fig. 14.23 (*c*) and (*d*).

Mechanical properties

Microhardness tests, carried out on the top, unreheated beads of the selected welds, and summarised in Fig. 14.24, show the extremely complex effect of nitrogen. Weld O, free from microalloying elements, gave a sharp increase in hardness with more than 150 ppm N, although the other welds of lower nitrogen content had similar hardness levels. The effect of nitrogen was very much reduced when 35 ppm Ti was present (weld W). With more titanium (weld X), the final addition of nitrogen actually reduced hardness. Addition of a small amount of boron (40 ppm, weld Y) resulted in a drop in hardness with the first nitrogen addition and then a smaller increase. With high boron contents, (welds U and T, with and without Ti) nitrogen was a softening element – particularly the second addition.

The tensile properties of all additional welds are given in Table 14.8; these show a complex pattern of behaviour and the strength properties of the selected welds are plotted in different ways in Fig. 14.25. The patterns of behaviour are roughly similar to the trends in hardness with nitrogen, already discussed in connection with Fig. 14.24, although the details are different. Boron can be seen in Fig. 14.26 to have reduced yield strength with increased levels of nitrogen, and at high boron content the normal nitrogen level of 80 ppm usually gave the strongest weld.

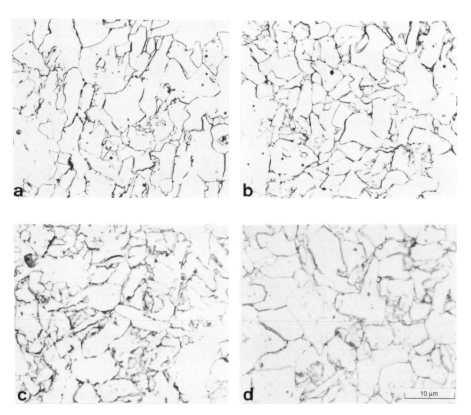

14.23 Microstructure of fine-grained reheated regions of welds with: *a*) 400 ppm Ti and 80 ppm N; *b*) 450 ppm Ti and 250 ppm N; *c*) 400 ppm Ti, 160 ppm B and 80 ppm N; *d*) 400 ppm Ti, 170 ppm B and 220 ppm N.

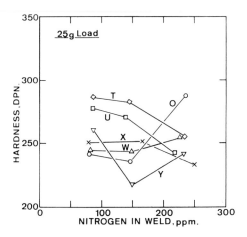

14.24 Influence of weld nitrogen content on top bead hardness of selected as-welded 1.4%Mn deposits with varying titanium and boron content.

Table 14.8 Mechanical properties of 1.4%Mn welds with varying boron, titanium and nitrogen content

N/Ti	B, ppm	Yield strength, N/mm²	Tensile strength, N/mm²	Elongation, %	Reduction of area, %	100 J Charpy transition, °C	
						AW	SR
160	2	510	570	25	76	−29	nd
120	32	530	580	26	76	−36	nd
	57	520	560	26	77	−37	nd
	90	480	550	27	75	−43	nd
	129	470	530	26	75	−46	nd
	158	450	520	27	76	−47	nd
160	2	550	610	24	78	−35	−58
250	46	510	570	26	75	−36	−46
	105	500	560	25	77	−43	−51
	147	500	550	24	77	−48	−56
	200	490	570	23	76	−50	−65
160	2	580	630	25	78	−44	−58
400	15	550	610	24	74	−48	−52
	29	530	590	27	74	−48	−57
	40	520	580	27	74	−48	−62
	54	520	580	27	75	−52	−58
	102	500	560	28	76	−64	−68
	158	490	560	23	73	−67	−68
240	2	520	600	29	77	−23	−54
120	29	530	590	28	76	−26	−62
	59	540	590	26	76	−22	−57
	90	520	580	26	71	−29	−50
	128	500	550	26	75	−34	−58
	162	470	530	29	75	−35	−61
240	2	550	620	24	76	−24	−46
250	43	530	590	24	73	−18	−35
	91	510	580	27	74	−22	−48
	143	490	560	27	74	−34	−47
	200	500	550	24	75	−46	−53
240	2	550	630	25	73	−30	−43
300	16	540	630	19	71	−30	−37
	28	540	610	24	71	−27	−40
	44	560	620	24	72	−24	−41
	52	540	600	25	70	−25	−41
	110	490	560	24	72	−38	−48
	167	470	540	28	73	−52	−52

Notes: for properties of welds with 80 ppm N see Table 13.8 and 14.4.
AW – as-welded; SR – stress relieved at 580 °C; nd – not determined.

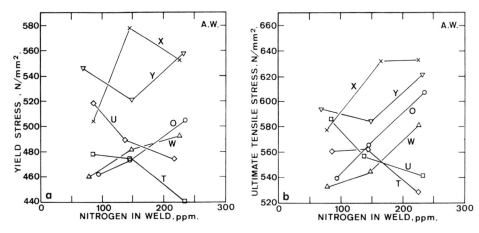

14.25 Influence of weld nitrogen content on: *a*) Yield strength; *b*) Tensile strength, of selected as-welded 1.4%Mn deposits with varying titanium and boron content.

When yield strength is plotted against boron content for the different titanium levels in Fig. 14.27, it can again be seen that, despite scatter, boron softened the weld metal if nitrogen exceeded about 80 ppm. Contrariwise, titanium appeared to strengthen weld deposits at 80 ppm nitrogen and all levels of boron. Its effect was much more complex at higher nitrogen levels and the positions of the <5 and 35 ppm Ti lines crossed, as did those of the higher titanium levels.

Much more work would be necessary to explain the results in detail but the trends described are consistent with boron combining with nitrogen so that the individual strengthening effects of both were nullified when they were present at relatively high levels.

The results of Charpy tests carried out on all welds, as-welded and after stress relief at 580 °C, are summarised as 100 J temperatures in Table 14.8 and Fig. 14.28–14.32. The detrimental effect of nitrogen on toughness in the as-welded condition is seen most clearly in Fig. 14.28. Only at the highest boron and titanium levels in Fig. 14.28 (*e*) did the transition temperature with the normal nitrogen level of 80 ppm rise above that of the deposits with 160 ppm N. Although stress relief greatly improved the toughness of welds with higher nitrogen contents, it only did so for the normal nitrogen content in the absence of titanium. Nevertheless, nitrogen was usually harmful to toughness in the PWHT condition.

In the as-welded condition at low titanium levels, boron was generally beneficial, although with more than 350 ppm Ti, excess boron

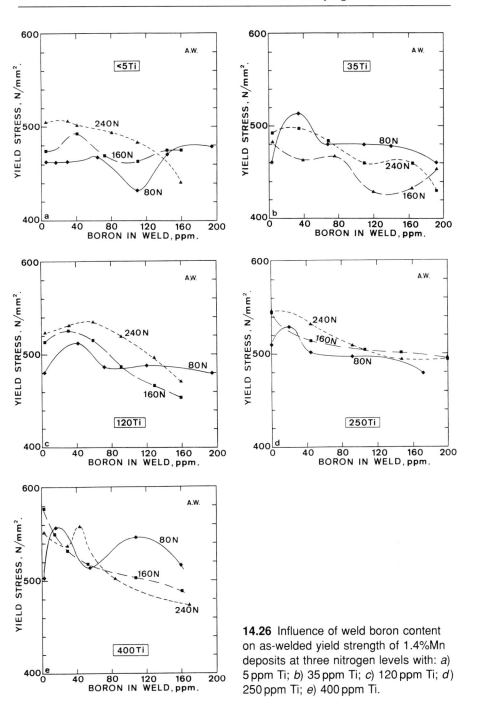

14.26 Influence of weld boron content on as-welded yield strength of 1.4%Mn deposits at three nitrogen levels with: *a*) 5 ppm Ti; *b*) 35 ppm Ti; *c*) 120 ppm Ti; *d*) 250 ppm Ti; *e*) 400 ppm Ti.

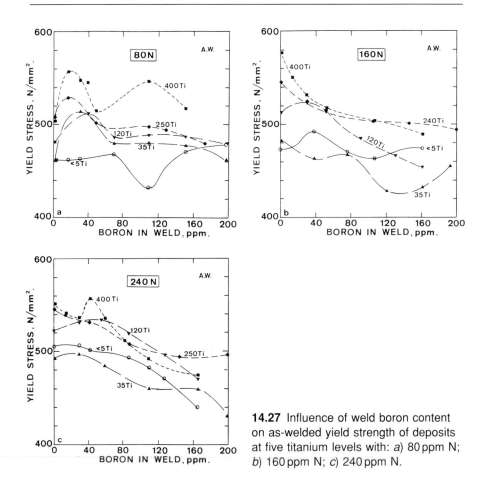

14.27 Influence of weld boron content on as-welded yield strength of deposits at five titanium levels with: *a*) 80 ppm N; *b*) 160 ppm N; *c*) 240 ppm N.

could lead to a deterioration in toughness. After stress relief, the role of boron appeared to be more complicated, although the effects were generally small. The beneficial effect of titanium in the as-welded condition at all nitrogen levels can be seen more clearly in Fig. 14.29. At the lowest nitrogen level (Fig. 14.29(*a*)), boron was beneficial with high titanium contents over a restricted range and with no added titanium, particularly above 80 ppm. However, with a nominal 35 ppm Ti addition boron was never beneficial. The best toughness (Charpy temperatures close to –80 °C) was obtained with either 20–60 ppm B and 400 ppm Ti or with 70–110 ppm B and 250 ppm Ti.

At increased nitrogen levels (Fig. 14.29(*b*) and (*c*)), the beneficial effect of boron was lost, except when boron and titanium were both high (a composition range not giving the best toughness when

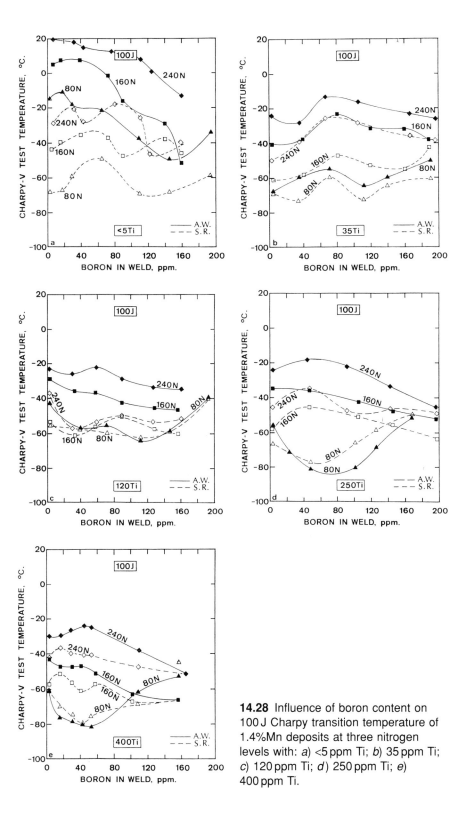

14.28 Influence of boron content on 100 J Charpy transition temperature of 1.4%Mn deposits at three nitrogen levels with: *a*) <5 ppm Ti; *b*) 35 ppm Ti; *c*) 120 ppm Ti; *d*) 250 ppm Ti; *e*) 400 ppm Ti.

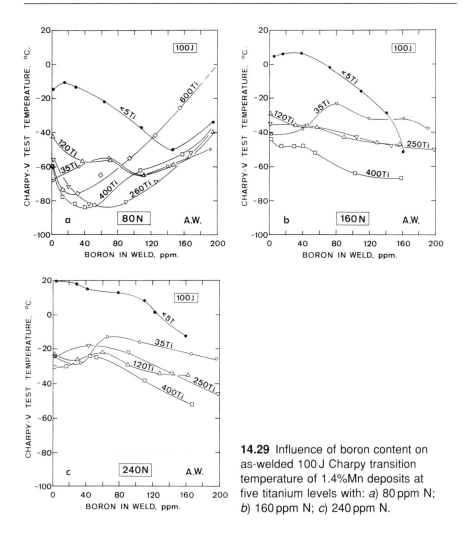

14.29 Influence of boron content on as-welded 100 J Charpy transition temperature of 1.4%Mn deposits at five titanium levels with: *a*) 80 ppm N; *b*) 160 ppm N; *c*) 240 ppm N.

nitrogen is lower) and also with no titanium and the maximum boron content. Deposits with the optimum boron and titanium levels described in the previous paragraph, therefore, showed little advantage over the standard composition with 35 ppm Ti without boron when severe nitrogen pick-up is likely. With heavy nitrogen contamination, the beneficial effect of boron was further reduced at all titanium levels.

After PWHT (Fig. 14.30), differences in composition had a smaller influence on toughness, partly because of the reduced influence of nitrogen and partly because of the improvement in toughness of the

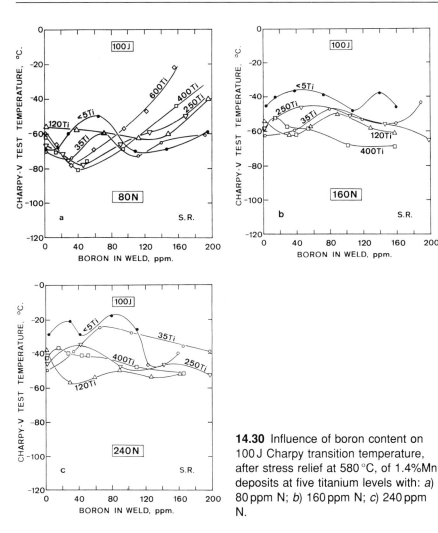

14.30 Influence of boron content on 100 J Charpy transition temperature, after stress relief at 580 °C, of 1.4%Mn deposits at five titanium levels with: a) 80 ppm N; b) 160 ppm N; c) 240 ppm N.

deposits free from titanium. Nevertheless, the lowest transition temperatures were found (as with the as-welded deposits) in welds of normal nitrogen content with 30–60 ppm B and 250–400 ppm Ti.

Three-dimensional plots of 100 J temperature against titanium and boron content at the three different nitrogen levels (Fig. 14.31 and 14.32), show a fuller picture. Both as-welded and stress-relieved, the valley at about 35 ppm Ti, which continued to the back of the diagram at the highest boron level, was flattened by nitrogen, as was the cross valley at about 40 ppm B, which deepened with increasing titanium content.

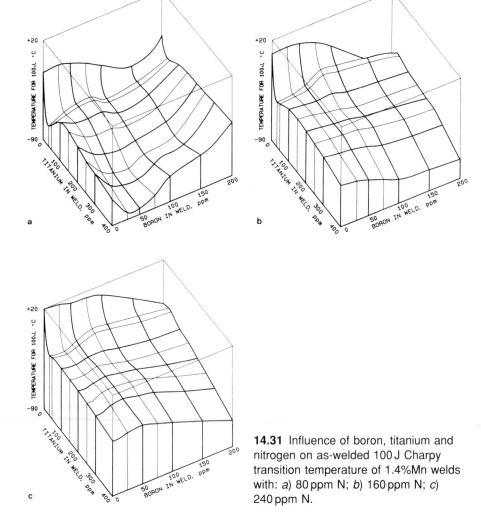

14.31 Influence of boron, titanium and nitrogen on as-welded 100 J Charpy transition temperature of 1.4%Mn welds with: a) 80 ppm N; b) 160 ppm N; c) 240 ppm N.

Titanium, boron, aluminium and nitrogen (Ti–B–Al–N)

To extend the work on 1.4%Mn welds with additions of aluminium, boron and titanium discussed earlier, further electrodes were made[9] to give welds with the two higher levels of nitrogen (N1, 160 ppm and N2, 240 ppm) considered in Chapter 12. The compositions of the additional welds are given in Table 14.9. Their relation to the matrix

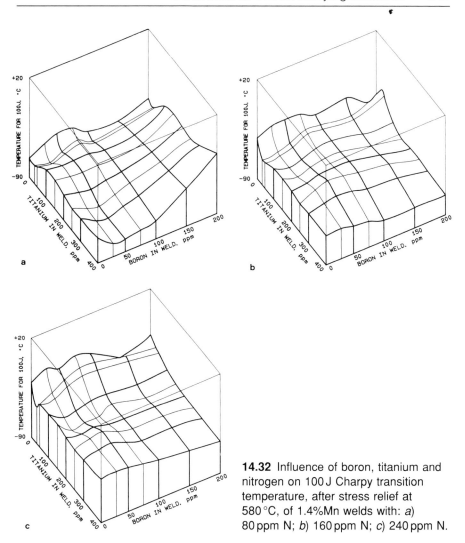

14.32 Influence of boron, titanium and nitrogen on 100 J Charpy transition temperature, after stress relief at 580 °C, of 1.4%Mn welds with: *a*) 80 ppm N; *b*) 160 ppm N; *c*) 240 ppm N.

of welds with T, B and N additions, given in Fig. 14.19, is shown in Fig. 14.33. Particular attention was paid to deposits coded as follows:

Y: 400 ppm Ti: 40 ppm B: 13 ppm Al
Z: 400 ppm Ti: 40 ppm B: 170 ppm Al
V: 400 ppm Ti: 40 ppm B: 570 ppm Al

Compositions were better balanced than in the deposits with 80 ppm N (Table 14.5) but oxygen and silicon both increased as aluminium was increased, as in the earlier welds.

Table 14.9 Composition of 1.4%Mn weld metal with titanium and boron having varying nitrogen and aluminium content

Element, wt%

C	Mn	Si	S	P	Ti	B	Al	N	O
0.069	1.48	0.34	0.007	0.010	0.037	0.0040	0.0005	0.015	0.028
0.070	1.49	0.34	0.007	0.010	0.041	0.0036	0.0061	0.015	0.038
0.070	1.45	0.43	0.006	0.010	0.047	0.0037	0.017	0.013	0.044
0.066	1.43	0.43	0.005	0.010	0.049	0.0041	0.033	0.012	0.049
0.069	1.44	0.55	0.005	0.010	0.046	0.0048	0.046	0.012	0.048
0.067	1.44	0.63	0.005	0.010	0.048	0.0044	0.056	0.012	0.047
0.069	1.51	0.36	0.007	0.008	0.041	0.0044	0.0005	0.023	0.029
0.067	1.38	0.35	0.007	0.012	0.050	0.0038	0.0062	0.025	0.041
0.068	1.45	0.50	0.006	0.011	0.047	0.0045	0.018	0.023	0.044
0.067	1.44	0.54	0.005	0.014	0.045	0.0042	0.031	0.023	0.048
0.069	1.38	0.52	0.006	0.011	0.041	0.0035	0.038	0.023	0.048
0.069	1.42	0.60	0.006	0.012	0.043	0.0035	0.056	0.024	0.047

Notes: all welds contained ~0.03%Cr, Ni and Cu, 0.005%Mo, <5 ppm Nb and V. For compositions of welds with 80 ppm N see Table 14.5.

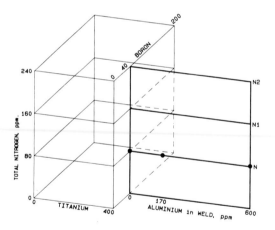

14.33 Matrix of experimental 1.4%Mn compositions with varying Ti, B and N, showing location of welds with variable aluminium content and coded welds selected for special consideration.

Metallography

Microstructural changes resulting from aluminium additions, shown in Fig. 14.34, may be compared with the changes found in welds with a nominal 80 ppm N shown in Fig. 14.14. Nitrogen reduced the amount of acicular ferrite in most deposits and tended to increase

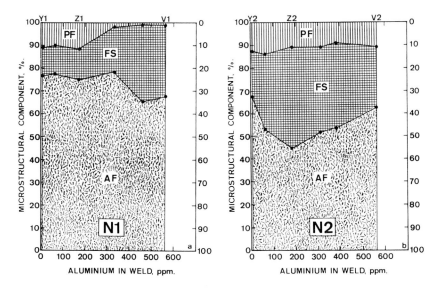

14.34 Influence of aluminium on the as-deposited microstructure of welds with 1.4%Mn, 400 ppm Ti and 40 ppm B at: *a*) 130 ppm N; *b*) 230 ppm N.

both ferrite with second phase and primary ferrite. However, at 230 ppm N, increasing aluminium contents above about 200 ppm increased the proportion of acicular ferrite at the expense of ferrite with second phase. The same effect was also found, to a smaller degree, between 460 and 560 ppm Al in 130 ppm N deposits. However, at lower aluminium contents the intermediate nitrogen system had behaved like the 80 ppm N deposits, giving a sharp increase in the amount of ferrite with aligned second phase and a decrease, almost to zero, of primary ferrite as aluminium was increased above 400 ppm.

Some of these changes are illustrated for welds of high nitrogen content in Fig. 14.35. With 230 ppm N and without added aluminium (Fig. 14.35(*a*)), a classical side-plate structure extended from a prior austenite boundary into acicular ferrite – a rare occurrence in the corresponding weld with 80 ppm N (Fig. 14.15(*a*)). Increasing aluminium to 180 ppm (Fig. 14.35(*b*)), led to a very coarse degenerate form of ferrite with second phase, still with appreciable primary and acicular ferrite. At the highest level of 560 ppm Al (Fig. 14.35(*c*)), the acicular ferrite was refined, with less ferrite with second phase but appreciable primary ferrite.

In the coarse-grained reheated microstructure, addition of nitrogen (Fig. 14.36) led to coarsening of the primary ferrite envelopes round

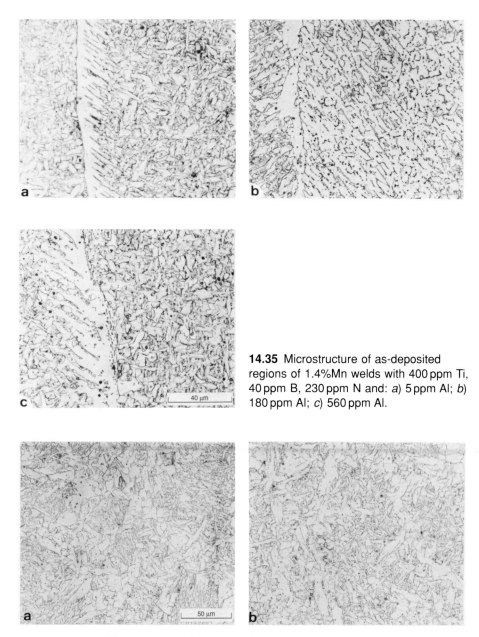

14.35 Microstructure of as-deposited regions of 1.4%Mn welds with 400 ppm Ti, 40 ppm B, 230 ppm N and: *a*) 5 ppm Al; *b*) 180 ppm Al; *c*) 560 ppm Al.

14.36 Microstructure of coarse-grained reheated regions of 1.4%Mn welds with 400 ppm Ti, 40 ppm B, 230 ppm N and: *a*) 180 ppm Al; *b*) 560 ppm Al.

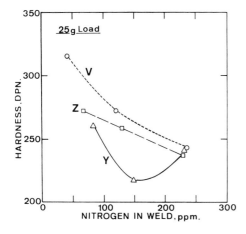

14.37 Influence of nitrogen on microhardness of as-welded deposits with 1.4%Mn, 400 ppm Ti and 40 ppm B at three levels of aluminium: Y 13 ppm, Z 170 ppm and V 570 ppm.

the prior austenite grains. Nitrogen also coarsened the interiors of the grains and led to microstructures having a high proportion of ferrite with second phase at the intermediate aluminium level, but it refined the grain interiors at the high aluminium level, replacing ferrite with second phase by acicular ferrite. In the fine-grained regions, nitrogen modified the microphase morphology and, in particular, coarsened the grain structure at the intermediate aluminium level.

Mechanical properties

Microhardness values of the as-deposited microstructure for the three selected welds, plotted against nitrogen content in Fig. 14.37, show that, whereas deposit Y with no added aluminium was softened and then hardened by increasing nitrogen, the presence of 170 and 570 ppm Al (welds Z and Y) caused nitrogen to produce softening, the softening being greater with the highest level of nitrogen.

Mechanical properties of deposits with enhanced nitrogen content are given in Table 14.10. Compared with its effect at the normal nitrogen level of 80 ppm (where aluminium caused strengthening), aluminium had very little influence on weld strength (Fig. 14.38), except for a slight increase in tensile strength with increasing aluminium at the 130 ppm N level. Plotted against nitrogen content, the strength of the three selected welds (Fig. 14.39), showed nitrogen to have little

Table 14.10 Mechanical properties of 1.4%Mn welds with titanium and boron having varying nitrogen and aluminium content

N, ppm	Al ppm	Yield strength, N/mm²	Tensile strength, N/mm²	Elongation, %	Reduction of area, %	100 J Charpy transition, °C	
						AW	SR
130	5	530	580	27	74	−48	−62
	61	560	590	25	69	−45	−48
	170	530	580	28	77	−43	−35
	330	540	600	27	73	−60	−56
	460	570	630	22	75	−56	−57
	560	570	640	25	74	−64	−55
230	5	540	600	24	72	−24	−41
	62	520	600	25	73	−4	−24
	180	490	580	27	69	13	−7
	310	520	600	24	72	−1	−18
	380	520	610	23	69	−14	−28
	560	530	590	26	74	−45	−45

Notes: for properties of welds with 80 ppm N see Table 14.6.
AW – as-welded; SR – stress-relieved at 580 °C.

effect on strength when no aluminium was added but gave apprecia-ble softening with intermediate levels of aluminium.

As with the strength properties, the 100 J Charpy temperature is plotted against aluminium in Fig. 14.40 and against nitrogen in Fig. 14.41. As-welded, adding nitrogen increased the transition tempera-ture of the basic welds, but the increased nitrogen content countered the harmful influence of aluminium above 160 ppm, so that above that level, aluminium tended to improve toughness at 130 ppm N and at 240 ppm N was wholly beneficial. Consequently, the 570 ppm Al weld with 240 ppm N was as tough as the zero aluminium weld with 130 ppm N and tougher than the 80 ppm N weld with an aluminium content of 570 ppm.

Stress-relieved (Fig. 14.40(*b*)), a similar pattern was evident; at increased nitrogen levels, aluminium up to 160 ppm was harmful but above that content was beneficial. At the highest aluminium level, the transition temperatures of the welds with enhanced nitrogen levels were similar to each other and lower than that of the 80 ppm N weld.

Plotting the Charpy temperature of the three selected welds V, Y and Z against nitrogen (Fig. 14.41), nitrogen was seen to be harmful

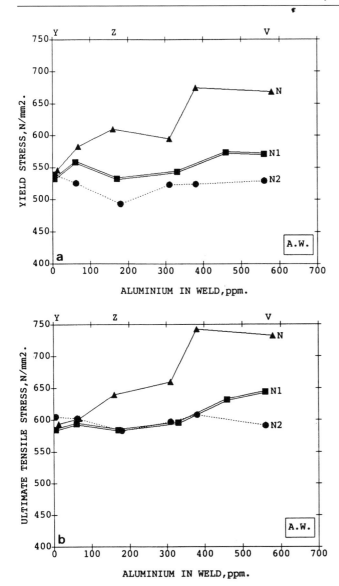

14.38 Influence of aluminium and nitrogen content on strength of as-welded deposits with 1.4%Mn, 400 ppm Ti and 40 ppm B. N, N₁ and N₂ refer to welds with 80, 160 and 240 ppm N, respectively.

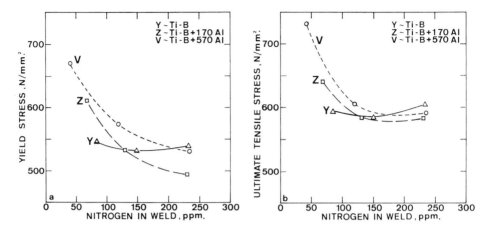

14.39 Influence of nitrogen on strength of as-welded deposits with 1.4%Mn, 400 ppm Ti and 40 ppm B at three levels of aluminium: Y 13 ppm, Z 170 ppm and V 570 ppm.

to the welds with zero and 70 ppm Al, whereas it is initially beneficial to the deposit of highest aluminium content in both as-welded and stress-relieved conditions. The damaging influence of nitrogen on toughness of deposits of low and intermediate aluminium content was more severe as-welded than stress-relieved and worse for the weld with 170 ppm Al, where a deterioration of ~100 °C on increasing nitrogen to the highest level was accompanied by a marked change from an acicular ferrite microstructure to a roughly equal mixture of acicular ferrite and ferrite with second phase.

General discussion

A comprehensive survey of effects of combinations of microalloying elements in a clean C–Mn system has shown that excellent combinations of toughness and strength could be obtained either with a very small addition (~35 ppm) of titanium or a larger addition of 300–400 ppm Ti. Both formulations give essentially acicular ferrite microstructures. The former system could not be improved by adding boron but an addition of ~40 ppm B improved toughness of the higher titanium welds and gave 100 J temperatures around –80 °C, over 10 °C lower than is possible with 35 ppm Ti alone.

Even with addition of aluminium, toughness in both systems deteriorated when nitrogen content was increased, although the detrimental influence of aluminium on toughness can be nullified if nitrogen content is high.

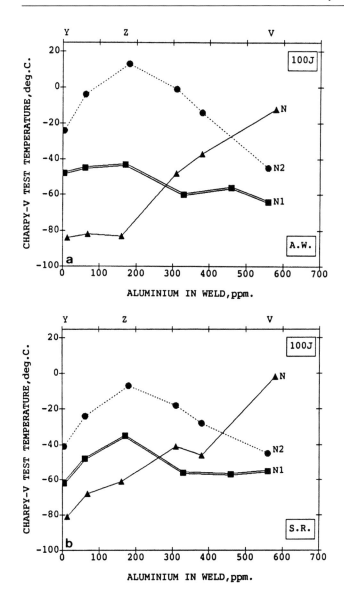

14.40 Influence of aluminium and nitrogen content on 100 J Charpy transition temperature of deposits with 1.4%Mn, 400 ppm Ti and 40 ppm B: a) As-welded; b) After stress relief at 580 °C.

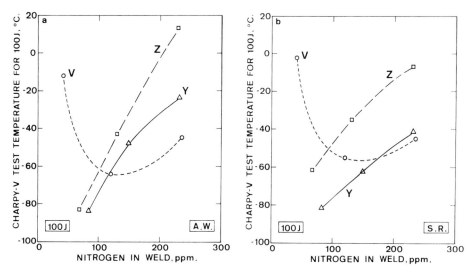

14.41 Influence of nitrogen on 100 J Charpy transition temperature of deposits with 1.4%Mn, 400 ppm Ti and 40 ppm B at three levels of aluminium: Y 13 ppm, Z 170 ppm and V 570 ppm: *a*) As-welded; *b*) After stress relief at 580 °C.

References

1 Evans GM: 'Microstructure and mechanical properties of ferritic steel welds containing Al and Ti'. *Weld J* 1995 **74**(8) 249s–61s; *Schweissmitteilungen* 1994 **52**(130) 21–39.

2 Tsuboi J and Terashima H: 'Review of strength and toughness of Ti and Ti–B microalloyed deposits'. *Welding in the World* 1983 **21**(11) 304–6.

3 Grong O and Matlock DK: 'Microstructural development in mild and low alloy steel weld metals'. *Int Met Rev* 1986 **31**(1) 27–48.

4 Oh DW, Olson DL and Frost RH: 'The influence of boron and titanium on low-carbon steel weld metal'. *Weld J* 1990 **69**(4) 151s–8s.

5 Mori N et al: 'The behaviour of B and N in notch toughness improvement of Ti–B-bearing weld metals'. IIW Doc IX–1158–80.

6 Ohkita S et al: 'The effect of oxide inclusions on microstructure of Ti–B containing weld metal'. *Australian Welding J* 1984 **28** 29–36.

7 Okaguchi S et al: 'Recent development of sour-gas service line pipe for North Sea use'. *The Sumitomo Search* 1990 (43) 25–34.

8 Evans GM: 'Microstructure and mechanical properties of ferritic steel welds containing Ti & B'. *Weld J* 1996 **75**(8) 251s–60s; IIW Doc II–A–932–94.

9 Evans GM: 'Effect of aluminium and nitrogen on Ti–B containing steel welds'. IIW Doc II–A–989–96.

Part V

Microalloying of high purity low alloy steel weld metals

15

Titanium and single additions of Cr, Mo, Ni, and Cu

The major alloying elements studied in Chapter 8 were deposited from electrodes formulated with raw materials which produced weld metals containing small amounts of microalloying elements. In this chapter, high purity weld metals containing the same major alloying elements were studied at 0.6 and 1.4%Mn, both with (code A and C) and without (code A_0 and C_0) the small addition of about 35 ppm Ti shown in Chapter 10 to be capable of nucleating acicular ferrite in C–Mn weld metals. Other microalloying elements were held at very low levels (<5 ppm). Separate additions of 1, 1.5 and 2%Cr; 0.25, 0.5 and 1%Mo; 1, 2.2 and 3%Ni and 0.5, 1 and 1.6%Cu were made and tested as-welded only. Experimental techniques were as used in other stages of the project, except that the sides of the weld preparation were buttered to minimise compositional differences due to dilution. Some of these results have not been published before.

Low manganese (without titanium)

Composition (Table 15.1) was well balanced in this, as in the other series. Carbon content was intentionally higher than that of welds in Chapter 8, averaging 0.075% compared with 0.046%; otherwise compositions were similar, apart from the lower levels of microalloying elements, and sulphur and phosphorus.

Mechanical properties

As-welded tensile and Charpy properties are given in Table 15.2. Generally (Fig 15.1), molybdenum was the most potent strengthening element, followed by Cr, Cu and Ni.

The first additions of all elements except chromium reduced 100 J Charpy temperature (Fig 15.2*). Subsequent additions (including manganese*) gave a sharp rise in transition temperature, except for

---

* In Fig. 15.2 and similar subsequent graphs, manganese in excess of the basic level (i.e. 0.6 or 1.4%) has been included in the graph.

Table 15.1 Composition of singly alloyed weld metal with 0.6%Mn, without titanium

Element, wt%	C	Mn	Si	S	P	Alloy	Ti	N	O
Addition, %									
None (A$_0$)	0.073	0.64	0.34	0.008	0.007	—	0.0007	0.006	0.045
Cr, 1	0.070	0.65	0.35	0.008	0.010	1.01Cr	0.0005	0.008	0.050
1.5	0.074	0.58	0.25	0.008	0.010	1.59	0.0004	0.009	0.050
2	0.076	0.64	0.28	0.008	0.008	2.10	0.0008	0.008	0.049
Mo, 0.25	0.070	0.66	0.31	0.006	0.008	0.24Mo	0.0005	0.008	0.043
0.5	0.073	0.66	0.34	0.007	0.007	0.50	0.0005	0.008	0.045
1	0.075	0.63	0.30	0.008	0.008	1.03	0.0005	0.008	0.044
Ni, 1	0.064	0.63	0.30	0.006	0.008	0.92Ni	0.0005	0.007	0.047
2.2	0.065	0.61	0.26	0.007	0.007	2.14	0.0005	0.009	0.049
3	0.062	0.62	0.29	0.007	0.007	3.01	0.0006	0.007	0.050
Cu, 0.5	0.068	0.62	0.28	0.008	0.007	0.50Cu	0.0005	0.007	0.048
1	0.065	0.62	0.30	0.007	0.009	1.03	0.0004	0.007	0.052
1.6	0.064	0.64	0.32	0.007	0.008	1.60	0.0004	0.006	0.050

Note: unless otherwise stated, all welds contained ~0.03%Ni, Cr, Cu, ~0.005%Mo, <5 ppm Al, B, Nb, V.

Table 15.2 As-welded mechanical properties of singly alloyed welds with 0.6%Mn without titanium

Addition, %	Yield strength, N/mm^2	Tensile strength, N/mm^2	Elongation, %	Reduction of area, %	Charpy transition, °C	
					100 J	28 J
None (A$_0$)	400	490	32	76	−46	−70
Cr, 1	480	560	22	75	−40	−68
1.5	490	590	19	72	−22	−48
2	560	660	nd	69	−26	−40
Mo, 0.25	430	520	26	79	−61	−87
0.5	470	560	25	75	−42	−76
1	530	620	21	73	7	−57
Ni, 1	420	500	32	77	−63	−88
2.2	440	530	32	76	−65	−95
3	460	540	28	77	−65	−98
Cu, 0.5	390	470	29	77	−63	−86
1	440	510	32	79	−54	−87
1.5	500	560	27	73	−24	−70

Note: nd – not determined.

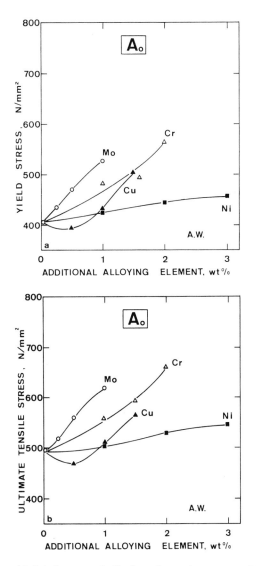

15.1 Influence of alloying elements on as-welded strength of high purity 0.6%Mn weld metal without titanium: *a*) Yield strength; *b*) Tensile strength.

nickel. At the 1% level, molybdenum and manganese gave larger increases than copper and chromium. However, even with nickel, the level of 100 J at just below −60 °C, with a yield strength no better than 460 N/mm², gave a combination of properties inferior to what is possible without major alloying but with 40 ppm Ti.

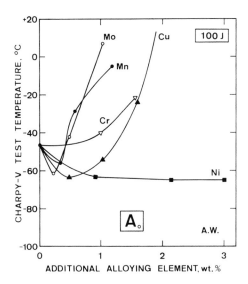

15.2 Influence of alloying elements on as-welded 100 J Charpy transition temperature of high purity 0.6%Mn weld metal without titanium.

Low manganese (with titanium)

The composition of the welds examined, given in Table 15.3, requires no comment. The titanium contents of 30–50 ppm were within the range where high acicular ferrite content and good toughness can be expected in C–1.4%Mn weld metals.

Mechanical properties

The as-welded mechanical properties (Table 15.4 and Fig. 15.3) show that, as with the titanium-free 0.6%Mn deposits, molybdenum provided the greatest strengthening and nickel the least. The 100 J Charpy temperature (Fig. 15.4) started from a similar level to the Ti-free welds but showed some divergences. The beneficial effect of nickel was more marked and manganese proved to be beneficial up to a 0.8% increase (i.e. a manganese content of 1.4%). The harmful effects of molybdenum and copper were lessened and were not apparent until 1% had been added. However, chromium had a greater (harmful) influence than without titanium.

Table 15.3 Composition of singly alloyed weld metal with 0.6%Mn and 35 ppm Ti

Element, wt%	C	Mn	Si	S	P	Alloy	Ti	N	O
Addition, %									
None (A)	0.073	0.64	0.35	0.005	0.004	—	0.0038	0.008	0.043
Cr, 1	0.077	0.66	0.32	0.008	0.010	1.03Cr	0.0034	0.008	0.046
1.5	0.078	0.62	0.26	0.007	0.005	1.60	0.0029	0.010	0.047
2	0.078	0.64	0.28	0.008	0.007	2.15	0.0042	0.010	0.048
Mo, 0.25	0.069	0.68	0.33	0.006	0.008	0.24Mo	0.0036	0.007	0.040
0.5	0.073	0.66	0.34	0.007	0.007	0.50	0.0042	0.007	0.043
1	0.072	0.66	0.32	0.007	0.007	1.03	0.0040	0.008	0.041
Ni, 1	0.063	0.65	0.31	0.006	0.009	0.88Ni	0.0038	0.009	0.047
2.2	0.064	0.61	0.27	0.007	0.007	2.17	0.0032	0.009	0.045
3	0.067	0.65	0.30	0.007	0.007	3.03	0.0036	0.007	0.046
Cu, 0.5	0.063	0.64	0.30	0.008	0.008	0.53Cu	0.0043	0.007	0.045
1	0.064	0.63	0.30	0.007	0.007	1.05	0.0042	0.006	0.046
1.6	0.066	0.67	0.37	0.006	0.008	1.44	0.0051	0.006	0.045

Note: unless otherwise stated, all welds contained ~0.03%Cr, Ni, Cu, ~0.005%Mo, <5 ppm Al, B, Nb, V.

Table 15.4 As-welded mechanical properties of singly alloyed welds with 0.6%Mn and 35 ppm Ti

Addition, %	Yield strength, N/mm^2	Tensile strength, N/mm^2	Elongation, %	Reduction of area, %	Charpy transition, °C	
					100 J	28 J
None (A)	400	490	33	78	−45	−70
Cr, 1	490	560	21	78	−35	−54
1.5	510	600	17	74	−8	−35
2	600	680	18	69	8	−33
Mo, 0.25	440	520	25	79	−58	−82
0.5	480	560	27	78	−48	−65
1	560	640	24	76	−33	−66
Ni, 1	420	500	32	77	−63	−88
2.2	460	540	28	79	−83	−113
3	480	560	29	78	−85	−114
Cu, 0.5	420	500	29	79	−62	−94
1	430	500	27	77	−58	−94
1.5	520	570	29	75	−43	−77

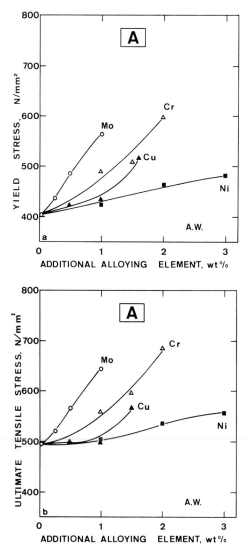

15.3 Influence of alloying elements on as-welded strength of high purity 0.6%Mn weld metal with 35 ppm Ti: *a*) Yield strength; *b*) Tensile strength.

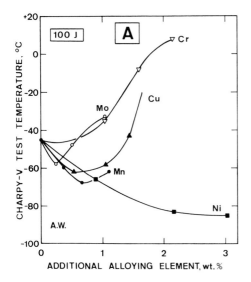

15.4 Influence of alloying elements on as-welded 100 J Charpy transition temperature of high purity 0.6%Mn weld metal with 35 ppm Ti.

Higher manganese (without titanium)

The composition of the welds with no titanium and 1.4%Mn (Table 15.5) shows that the elements were held to the desired levels.

Mechanical properties

Table 15.6 lists the as-welded mechanical properties, from which it appears that the higher manganese content increased strength and increased the strengthening effect of nickel (Fig. 15.5) but did not change the relative effects of the alloying elements.

At the higher manganese level nickel ceased to improve toughness and the first additions of any alloying element no longer improved toughness – although they did not impair it (Fig. 15.6). The maximum additions of each element (aside from Mn and Ni) were equally detrimental.

Higher manganese (with titanium)

The compositions of the welds examined are given in Table 15.7.

Table 15.5 Composition of singly alloyed weld metal with 1.4%Mn, without titanium

Element, wt%	C	Mn	Si	S	P	Alloy	Ti	N	O
Addition, %									
None (C$_0$)	0.078	1.43	0.26	0.006	0.006	—	0.0005	0.009	0.048
Cr, 1	0.080	1.48	0.28	0.007	0.006	1.07Cr	0.0005	0.009	0.044
1.5	0.075	1.48	0.28	0.007	0.005	1.55	0.0005	0.009	0.047
2	0.076	1.44	0.28	0.007	0.007	1.96	0.0005	0.008	0.048
Mo, 0.25	0.076	1.53	0.28	0.008	0.006	0.25Mo	0.0005	0.008	0.044
0.5	0.076	1.45	0.25	0.008	0.006	0.50	0.0005	0.008	0.046
1	0.076	1.41	0.24	0.008	0.006	1.11	0.0005	0.008	0.045
Ni, 1	0.073	1.39	0.28	0.007	0.007	0.95Ni	0.0005	0.008	0.045
2.2	0.071	1.48	0.26	0.007	0.012	2.28	0.0005	0.009	0.046
3	0.073	1.43	0.24	0.008	0.006	3.12	0.0005	0.008	0.046
Cu, 0.5	0.080	1.47	0.24	0.007	0.007	0.55Cu	0.0005	0.010	0.046
1	0.075	1.41	0.24	0.006	0.008	1.06	0.0005	0.009	0.048
1.6	0.074	1.36	0.22	0.008	0.010	1.63	0.0005	0.011	0.048

Note: unless otherwise stated, all welds contained ~0.03%Cr, Ni, Cu, ~0.005%Mo, <5 ppm Al, B, Nb, V.

Table 15.6 As-welded mechanical properties of singly alloyed welds with 1.4%Mn, without titanium

Addition, %	Yield strength, N/mm^2	Tensile strength, N/mm^2	Elongation, %	Reduction of area, %	Charpy transition, °C	
					100 J	28 J
None (C$_0$)	400	490	32	76	−46	−70
Cr, 1	570	660	18	69	−20	−70
1.5	590	690	18	70	6	−55
2	640	730	18	69	24	−20
Mo, 0.25	470	570	20	77	−24	−61
0.5	520	610	19	75	−4	−50
1	590	700	19	69	26	−50
Ni, 1	470	550	26	78	−13	−48
2.2	510	600	22	75	−20	−61
3	570	670	16	71	−12	−64
Cu, 0.5	450	540	27	79	−10	−36
1	490	570	25	75	−4	−48
1.5	560	640	23	71	23	−30

Mechanical properties

Titanium raised the general level of strength (Table 15.8) and increased still further the strengthening effect of nickel (Fig. 15.7), although the relative potency of the four elements was unchanged.

Table 15.7 Composition of singly alloyed weld metal with 1.4%Mn and 35 ppm Ti

Element, wt%	C	Mn	Si	S	P	Alloy	Ti	N	O
Addition, %									
None (C)	0.074	1.46	0.28	0.008	0.008	—	0.0035	0.008	0.044
Cr, 1	0.079	1.56	0.30	0.007	0.011	1.03Cr	0.0027	0.010	0.043
1.5	0.073	1.53	0.29	0.007	0.010	1.53	0.0027	0.008	0.044
2	0.078	1.48	0.29	0.007	0.010	1.96	0.0028	0.008	0.044
Mo, 0.25	0.076	1.49	0.26	0.008	0.008	0.25Mo	0.0034	0.009	0.041
0.5	0.075	1.46	0.26	0.008	0.007	0.50	0.0031	0.009	0.043
1	0.075	1.44	0.25	0.008	0.008	1.07	0.0031	0.008	0.043
Ni, 1	0.076	1.48	0.25	0.008	0.006	0.95Ni	0.0033	0.008	0.042
2.2	0.068	1.46	0.27	0.008	0.007	2.24	0.0034	0.008	0.044
3	0.075	1.46	0.26	0.008	0.006	3.15	0.0032	0.007	0.042
Cu, 0.5	0.076	1.48	0.27	0.007	0.008	0.53Cu	0.0022	0.008	0.045
1	0.077	1.43	0.25	0.007	0.008	1.08	0.0019	0.009	0.046
1.6	0.069	1.37	0.25	0.007	0.009	1.60	0.0023	0.008	0.048

Note: unless otherwise stated, all welds contained ~0.03%Cr, Ni, Cu, ~0.005%Mo, <5 ppm Al, B, Nb, V.

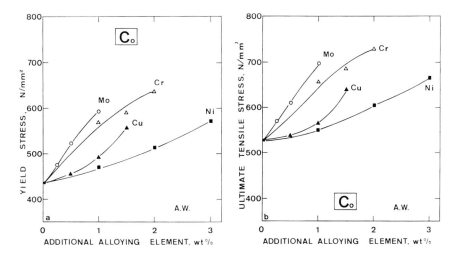

15.5 Influence of alloying elements on as-welded strength of high purity 1.4%Mn weld metal without titanium: *a)* Yield strength; *b)* Tensile strength.

Titanium also improved the basic level of toughness (Fig. 15.8) considerably and, at the same time, it reduced the harmful effects of molybdenum and manganese, although not of the other elements. Nickel, in fact, was found to be definitely harmful in this context; the maximum addition of nearly 3% raised the 100 J Charpy temperature from nearly −70 °C to approximately −50 °C.

Table 15.8 As-welded mechanical properties of singly alloyed welds with 1.4%Mn and 35 ppm Ti

Addition, %	Yield strength, N/mm²	Tensile strength, N/mm²	Elongation, %	Reduction of area, %	Charpy transition, °C	
					100 J	28 J
None (C)	460	530	28	79	−68	−88
Cr, 1	620	670	20	73	−58	−84
1.5	650	710	18	73	−38	−79
2	710	770	16	68	−6	−70
Mo, 0.25	560	600	25	77	−69	−86
0.5	600	640	21	74	−56	−90
1	650	720	20	71	−39	−85
Ni, 1	500	550	27	79	−62	−94
2.2	560	620	26	78	−58	−102
3	600	680	20	72	−49	−95
Cu, 0.5	500	560	28	79	−57	−93
1	490	560	27	77	−48	−85
1.5	570	620	26	75	−15	−67

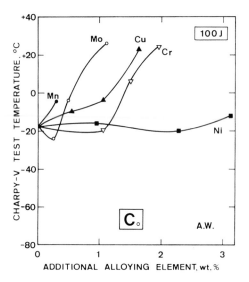

15.6 Influence of alloying elements on as-welded 100 J Charpy transition temperature of high purity 1.4%Mn weld metal without titanium.

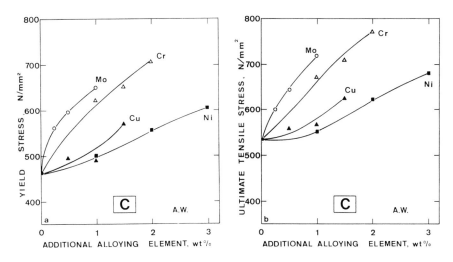

15.7 Influence of alloying elements on as-welded strength of high purity 1.4%Mn weld metal with 35 ppm Ti: a) Yield Strength; b) Tensile strength.

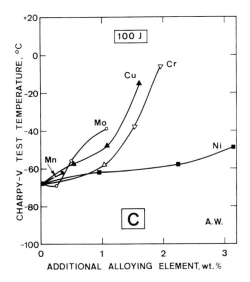

15.8 Influence of alloying elements on as-welded 100 J Charpy transition temperature of high purity 1.4%Mn weld metal with 35 ppm Ti.

Discussion

Although no supporting metallography was carried out, it is likely that ~40 ppm Ti has little effect on strength or toughness of alloyed pure weld metals unless they contain sufficient manganese and/or other alloying elements to give acicular ferrite in their microstructure. The significance of observations of acicular ferrite should be recognised. Acicular ferrite content is measured on as-deposited microstructure (which exists only to a limited extent in the regions from which test specimens were extracted). However, acicular ferrite in the as-deposited microstructure is a sure sign of its presence in the coarse-grained reheated structure. It is also a likely sign that the low temperature microstructure is refined and contains less harmful grain boundary cementite.

For chromium alloying, which produces a bainitic microstructure, toughness in the 0.6%Mn deposits was independent of the presence of titanium. With 1.4%Mn, the transition temperatures of deposits with and without titanium converged as the amount of alloying was increased.

Molybdenum, like manganese, promotes acicular ferrite, so that increasing alloying led to increased separation of toughness levels in deposits with and without titanium.

Nickel produced an intermediate state of affairs. With 0.6%Mn, welds with titanium continued improving in toughness as nickel was increased up to almost 3%, whereas without titanium, adding more than 1%Ni gave little improvement. However, with 1.4%Mn, as little as 1%Ni reduced toughness, more so with titanium than without it. Overall, however, in weld metals giving a useful level of strength, titanium was helpful in achieving the best toughness and the purer weld metals were generally tougher than the similar, less pure compositions examined in Chapter 8.

Copper, usually being an impurity in weld metals, was not examined in this way but may be regarded as relatively harmless if kept below 0.5%.

For higher strength and good toughness, weld metals of the Mn–Ni–Mo type, i.e. containing the two more promising elements discussed above, were explored. The results of tests on this type of weld metal are presented in the next chapter.

16

Titanium and multiple alloying elements: Mn–Ni–Mo weld metals

The experimental work described in this chapter was conducted by MQ Johnson as part of a doctoral thesis. The aim was to examine the influence of titanium on a more complex alloyed weld metal (Ref. 8, Ch. 4 and Ref. 1–3, this chapter) – a Mn–Ni–Mo type – capable of producing as-welded yield strength levels up to 900 N/mm² with good toughness. To achieve such high strength levels, restrictions are placed on the maximum interpass temperature and heat input although, in practice, these are not much more onerous than were applied in much of the work already described. In common with other alloyed welds, plates were buttered before welding. Interpass temperature and heat input were sometimes varied from the standard values of 150 °C and 1 kJ/mm, respectively, as indicated in the tables. With the exception of making the electrodes and welds and conducting the initial mechanical property testing, the work was carried out at the Colorado School of Mines, Golden, Co, USA.

Three series of welds were deposited, all with a nominal composition (wt%) of 0.04C–1.5Mn–3.2Ni–0.5Mo but with titanium varying between <10 and 1000 ppm. Other elements were as low, or as constant, as practicable, bearing in mind the influence of added metallic titanium and heat input on elements such as carbon, silicon and oxygen. The three series of welds were deposited with heat inputs of 1, 2.2 and 4.3 kJ/mm. In the first series only, the interpass temperature was varied between 25 and 300 °C, but in the other two it was either 100 or 150 °C. Two types of weld were deposited at 2.2 kJ/mm. One was of the ISO 2560:1973 type three beads wide, used previously, with Charpy notches located at the bead centres, to maximise the proportion of coarse-grained weld metal in the microstructure of the area tested. The other was of the AWS A5.1–69 type, two beads wide, with Charpy notches located at the bead overlap position, so as to maximise fine-grained weld metal. Otherwise, test techniques were similar to those reported elsewhere in this monograph.

Table 16.1 Composition of Mn–Ni–Mo weld metal having varying titanium content deposited with different interpass temperature: *a*) At 1 kJ/mm

Element wt%	C	Mn	Si	Ni	Mo	Ti	N	O
Interpass temp, °C								
25	0.043	1.64	0.35	3.45	0.47	0.0007	0.006	0.040
	0.050	1.61	0.35	3.30	0.47	0.0040	0.010	0.038
	0.054	1.48	0.29	3.31	0.44	0.0091	0.009	0.034
	0.051	1.59	0.39	3.38	0.46	0.025	0.008	0.023
	0.055	1.67	0.40	3.40	0.47	0.053	0.007	0.026
150	0.043	1.58	0.32	3.56	0.46	0.0007	0.008	0.041
	0.044	1.60	0.34	3.29	0.46	0.0042	0.008	0.034
	0.045	1.58	0.30	3.45	0.46	0.0095	0.009	0.027
	0.051	1.49	0.33	3.41	0.46	0.020	0.009	0.027
	0.056	1.65	0.38	3.38	0.46	0.047	0.006	0.025
150	0.049	1.60	0.30	3.32	0.45	0.0006	0.007	0.052
	0.047	1.59	0.32	3.26	0.45	0.0038	0.008	0.044
	0.050	1.66	0.33	3.27	0.46	0.0083	0.008	0.043
	0.052	1.67	0.32	3.31	0.48	0.012	0.008	0.038
	0.051	1.62	0.36	3.24	0.47	0.015	0.008	0.035
	0.054	1.55	0.32	3.54	0.47	0.023	0.008	0.029
	0.059	1.66	0.44	3.26	0.47	0.056	0.009	0.026
	0.066	1.66	0.43	3.22	0.47	0.100	0.006	0.025
240	0.043	1.62	0.31	3.44	0.47	0.0007	0.008	0.040
	0.043	1.64	0.34	3.33	0.47	0.0046	0.008	0.035
	0.046	1.43	0.25	3.31	0.44	0.0077	0.007	0.028
	0.053	1.52	0.33	3.35	0.46	0.022	0.007	0.027
	0.054	1.72	0.40	3.44	0.46	0.058	0.007	0.025
300	0.040	1.55	0.30	3.28	0.47	0.0007	0.006	0.041
	0.041	1.58	0.31	3.37	0.47	0.0041	0.006	0.039
	0.043	1.53	0.30	3.39	0.48	0.090	0.006	0.033
	0.049	1.53	0.37	3.34	0.46	0.024	0.008	0.028
	0.053	1.63	0.38	3.28	0.48	0.048	0.007	0.025

Weld analysis, detailed in Table 16.1, confirmed the consistency of the compositions, with the expected changes in content of carbon, silicon and oxygen. The oxygen content levelled out to about ~0.025%O when 250 ppm Ti was present (Fig. 16.1); this level was rather lower than the ~0.029%O found with simple C–Mn weld metals at similar titanium levels in Chapters 10 and 11 (see also Fig. 14.1). In addition, increasing heat input increased nitrogen content from ~70 to ~90 to ~120 ppm N at the three heat inputs examined. Such a relationship was not examined in Chapter 4, where the influence of heat input on C–Mn weld metals was evaluated.

Table 16.1   cont. b) Deposited at 2.2 kJ/mm

Element, wt%	C	Mn	Si	Ni	Mo	Ti	N	O
Interpass temp, °C								
100	0.042	1.51	0.28	3.33	0.47	<0.0010	0.008	0.051
AWS	0.041	1.54	0.29	3.34	0.48	0.0030	0.006	0.046
	0.043	1.55	0.27	3.32	0.48	0.0057	0.008	0.043
	0.045	1.50	0.26	3.32	0.48	0.0104	0.007	0.041
	0.053	1.51	0.33	3.33	0.46	0.0175	0.009	0.034
	0.058	1.55	0.31	3.32	0.48	0.0271	0.009	0.028
150	0.043	1.53	0.27	3.22	0.45	0.0005	0.010	0.045
	0.043	1.51	0.28	3.29	0.45	0.0030	0.010	0.039
	0.044	1.55	0.28	3.21	0.46	0.0064	0.009	0.037
	0.046	1.49	0.27	3.26	0.47	0.0095	0.010	0.036
	0.047	1.54	0.32	3.26	0.46	0.013	0.010	0.034
	0.050	1.41	0.25	3.73	0.47	0.018	0.010	0.030
	0.057	1.65	0.37	3.25	0.46	0.040	0.010	0.026
	0.063	1.62	0.41	3.27	0.47	0.079	0.010	0.020

Table 16.1   cont. c) Deposited at 4.3 kJ/mm

Element, wt%	C	Mn	Si	Ni	Mo	Ti	N	O
Interpass temp, °C								
150	0.039	1.64	0.25	3.33	0.47	0.0005	0.010	0.039
	0.044	1.51	0.24	3.18	0.46	0.0023	0.013	0.036
	0.043	1.66	0.24	3.21	0.47	0.0058	0.013	0.032
	0.046	1.53	0.23	3.12	0.44	0.0073	0.012	0.033
	0.047	1.56	0.30	3.14	0.44	0.010	0.012	0.030
	0.050	1.48	0.22	3.57	0.46	0.014	0.011	0.026
	0.055	1.48	0.31	3.51	0.45	0.032	0.014	0.021
	0.062	1.50	0.29	3.25	0.48	0.055	0.014	0.019

Note: all welds contained 0.003–0.008%S, 0.003–0.015%P, <20 ppm Al, <3 ppm B, ≤22 ppm V, <15 ppm Nb, ≤0.036%Cr, <0.035%Cu.

Metallography

As-deposited microstructures of welds deposited at 1 kJ/mm and 150 °C interpass temperature were influenced by titanium as shown in Fig. 16.2. Without titanium, ferrite with second phase formed nearly 95% of the microstructure, the remainder being acicular ferrite, as the amount of alloying was too great for primary ferrite to

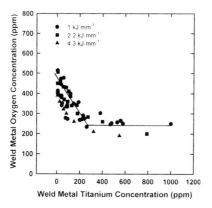

16.1 Effect of weld titanium on weld oxygen content for Mn–Ni–Mo weld metal deposited at three heat inputs.

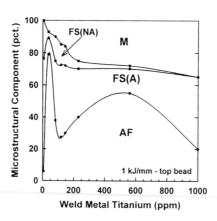

16.2 Effect of titanium content on as-deposited microstructure of Mn–Ni–Mo weld metal deposited at a heat input of 1 kJ/mm with 150 °C interpass temperature (AF – acicular ferrite, FS(A) – ferrite with aligned second phase, FS(NA) – ferrite with non-aligned second phase, M – martensite).

be present. As with C–Mn deposits, as little as 40 ppm Ti (the lowest addition made) changed the microstructure dramatically and gave a predominantly acicular ferrite (~80%) weld metal with a few percent of martensite and <20% ferrite with second phase. The martensite fraction continuously increased to nearly 40% as titanium increased to the maximum addition of 1000 ppm (0.1%), but the other constituents changed markedly. Acicular ferrite fell to a minimum of less than 30% at 120 ppm Ti, increased to nearly 60% at about 560 ppm Ti and then fell to ~20% at the maximum titanium addition. Ferrite

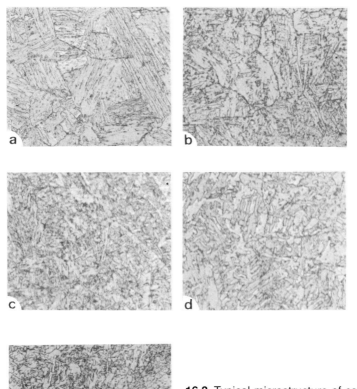

16.3 Typical microstructure of coarse-grained reheated Mn–Ni–Mo weld metal deposited at 1 kJ/mm with different interpass temperature: a) 7 ppm Ti, 25 °C; b) 7 ppm Ti, 150 °C; c) 40 ppm Ti, 25 °C; d) 40 ppm Ti, 150 °C; e) 470 ppm Ti, 150 °C.

with second phase (mostly aligned) occupied the remainder of the microstructure and was at another minimum at 560 ppm Ti. The changes in proportion of acicular ferrite are very similar to those in the simpler C–1.4%Mn–Ti weld metals detailed in Chapter 10.

The coarse-grained microstructure of titanium-free welds was ferrite with aligned second phase at higher cooling rates. With the rapid cooling induced by an interpass temperature of 25 °C, colonies of ferrite with aligned second phase spanned entire prior austenite grains (Fig. 16.3(a)). With slower cooling, polygonal ferrite (sometimes known as 'granular bainite') was introduced (Fig. 16.3(b)), and the aligned structures were smaller than the prior austenite grains. With 40 ppm Ti, acicular ferrite was introduced (Fig. 16.3(c) and

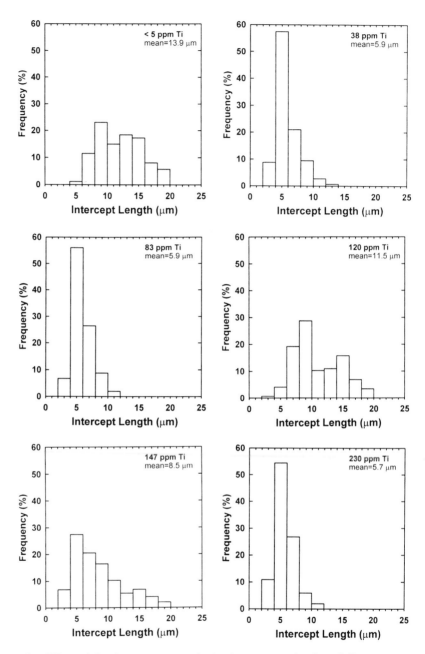

16.4 Effect of titanium content on ferrite intercept grain size of Charpy notch regions of Mn–Ni–Mo welds deposited at a heat input of 1 kJ/mm with 150 °C interpass temperature, Ti content shown on figures.

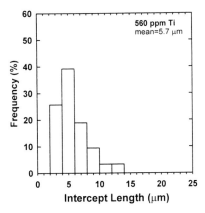

16.4 *(cont.)*

(*d*)); it was present at the highest and lowest cooling speeds, although martensite predominated with fast cooling. Generally, the coarse-grained reheated microstructures of the deposits with titanium reflected the as-deposited structures, although with high titanium levels (Fig. 16.3(*e*)), inhomogeneous banding and precipitation along prior austenite boundaries were both apparent.

Measurements of ferrite grain size in the regions sampled by Charpy notches (i.e. a mixture of coarse- and fine-grained reheated weld metal) from welds deposited at 1 kJ/mm with 150 °C interpass temperature are summarised in Fig. 16.4. The titanium-free deposit exhibited a broad spread of values, averaging 14 μm with no clear most frequently observed value. Other deposits have clear most frequent values of 5 μm. However, whereas welds with 38, 83, 230 and 560 ppm Ti all gave a mean grain size of about 5.7 μm, welds with 120 and 147 ppm Ti were coarser (8.5 and 11.5 μm, mean) and with maximum observed sizes of 20 μm, i.e. as coarse as for the deposit with no added titanium. These observations are consistent with acicular ferrite in the coarse-grained reheated microstructure being as fine as the fine-grained reheated microstructure and the other constituents being coarser.

A summary of an examination of inclusions present in selected examples of these weld metals is presented in Chapter 17.

Mechanical properties

Figure 16.5, taken from Ref. 16.1, shows that titanium generally had a strengthening effect, although this was somewhat complex and

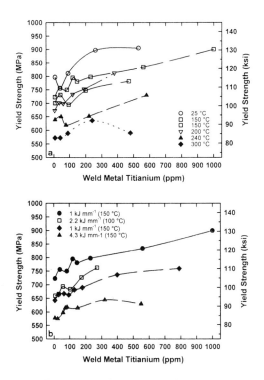

16.5 Effect on yield strength of Mn–Ni–Mo welds of titanium content and: a) Interpass temperature; b) Heat input.

possibly enhanced by minor changes in composition brought about by the titanium additions. Increasing interpass temperature weakened the weld metal, particularly above 200 °C (Fig. 16.5(a)). Increasing heat input (Fig. 16.5(b)), had a similar effect, which could be partially offset by reducing the interpass temperature.

The influence of titanium on as-welded 100 J Charpy temperature is shown in relation to interpass temperature in Fig. 16.6(a) and to heat input in Fig. 16.6(b). Both graphs reveal an improvement in toughness with the first addition of titanium, followed by an impairment at ~100 ppm Ti and an improvement at 200–300 ppm Ti, followed by a further worsening. The shapes of the curves approximately reverse that of acicular ferrite in Fig. 16.2 and resemble the behaviour of the simple 1.4%Mn weld metals with varying titanium content described in Chapter 10.

The best toughness was generally achieved with 150 °C interpass temperature and the worst with either 25 or 300 °C. Except in the absence of added titanium, varying heat input from 1 to 2.2 kJ/mm

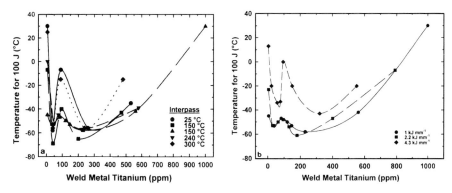

16.6 Effect on 100 J Charpy transition temperature of Mn–Ni–Mo welds of titanium content and: *a)* Interpass temperature; *b)* Heat input.

had little effect on toughness (Fig. 16.6(*b*)); increasing to 4.3 kJ/mm was always the least tough option.

Simulation experiments

Tests were made[3] to ascertain which of the regions in multipass welds were 'weak links' in the system and gave the lowest toughness. Simulations were made on single beads which had been deposited with a heat input of 2.2 kJ/mm into gouged-out multipass welds (Fig. 10.19) using an interpass temperature of either 25 °C to give a fine initial

Table 16.2 Composition of Mn–Ni–Mo weld metal having varying titanium content deposited with different interpass temperatures for simulation tests

Element, wt%		C	Mn	Si	Ni	Mo	Ti	N	O
Interpass temp, °C	Electrode code								
25	F1	0.040	1.51	0.26	3.35	0.46	0.0001	0.005	0.046
	F30	0.037	1.50	0.27	3.30	0.46	0.0030	0.005	0.034
	F100	0.039	1.56	0.27	3.02	0.45	0.010	0.006	0.031
	F250	0.045	1.49	0.34	3.09	0.46	0.025	0.006	0.023
	F420	0.047	1.58	0.34	3.00	0.43	0.042	0.006	0.022
240	C1	0.035	1.52	0.27	3.25	0.43	0.0001	0.007	0.041
	C30	0.038	1.51	0.28	3.35	0.46	0.0030	0.006	0.035
	C110	0.037	1.39	0.27	3.01	0.43	0.011	0.006	0.032
	C235	0.046	1.53	0.35	3.10	0.45	0.024	0.006	0.022
	C320	0.048	1.56	0.33	3.08	0.45	0.032	0.006	0.020

Note: all welds contained 0.005–0.007%S, 0.009–0.011%P, <20 ppm Al, <3 ppm B, ≤22 ppm V, <15 ppm Nb, ≤0.036%Cr, <0.035%Cu.

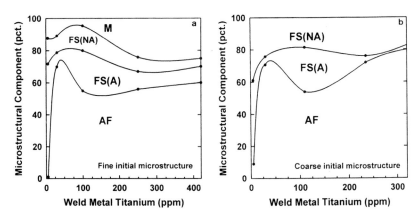

16.7 Effect of titanium content on as-deposited microstructure of Mn–Ni–Mo weld metal for simulation tests, deposited at a heat input of 2.2 kJ/mm with: *a*) 25 °C interpass; *b*) 240 °C interpass temperature.

microstructure, or 240 °C to give a coarse. The composition of these welds, which had titanium content ranging from 1 to 420 ppm, is given in Table 16.2.

A preliminary dilatometric study showed the A_{c1} and A_{c3} temperatures were 720 and 910 °C, respectively. Simulation was carried out in a Gleeble weld simulator, in a manner similar to that described in Chapter 10 (Fig. 10.19 and 10.20), except that sub-sized 10 × 5 mm Charpy specimens were extracted from 5 mm thick blanks, thermally simulated and tested at −80 °C. Metallography was carried out on selected specimens. Simulation temperatures selected were as follows:

850 °C, representing intercritically reheated region;
1050 °C, representing fine-grained reheated region;
1150 °C, representing fine-grained reheated region;
1350 °C, representing coarse-grained reheated region;
1350 °C + 850 °C, representing-coarse grained reheated material intercritically reheated.

Quantitative metallography on the as-welded 3.2 kJ/mm deposits plotted against titanium content (Fig. 16.7), revealed some differences from the results in Fig. 16.2, which were obtained at 1 kJ/mm with 150 °C interpass temperature (and were extended to higher titanium contents). Martensite was absent from all deposits made using a 240 °C interpass temperature and, although the amount of acicular ferrite was generally less, the very sharp minimum at 100–150 ppm

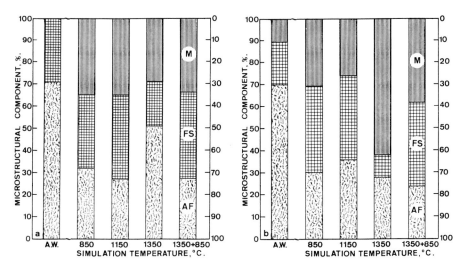

16.8 Microstructure of Mn–Ni–Mo–30 ppm Ti specimens simulated as indicated from: *a*) Fine-grained; *b*) Coarse-grained weld metal.

Ti in Fig. 16.2 was less severe when the higher heat input was employed.

Use of quantitative metallography of simulated specimens has so far been restricted to deposits with 30 ppm Ti, and the results (combining together ferrite with aligned and with non-aligned second phase) are compared with as-deposited values in Fig. 16.8. Although the starting points differed, the values for coarse and fine grained simulated welds differed by no more than about 5% except for the coarse grained specimens heated to 1350 °C. Here, the initially fine grained version contained over 60% martensite and the coarse grained one less than 30%. In general, simulation increased the proportion of martensite from the as-deposited level and also, except for the coarse-grained (1350 °C) simulation, reduced the proportion of acicular ferrite below that of ferrite with second phase. Even with the 1350 °C simulation, the proportion of acicular ferrite to ferrite with second phase was lower than in as-deposited weld metal.

Hardness and Charpy tests have been carried out at all titanium levels. Hardness (Fig. 16.9) increased with titanium level, the maximum values being at ~240 ppm Ti. All simulated deposits were harder than the as-welded (2.2 kJ/mm) levels, there being little dependence on the simulation treatment, except that the 1350 + 850 °C simulation softened the initially coarse deposit – but not the fine one (which had a much higher proportion of martensite).

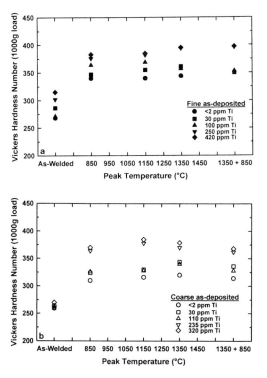

16.9 Effect of simulation treatment and titanium content on hardness of Mn–Ni–Mo weld metal: a) Originally fine-grained; b) Originally coarse grained.

Charpy results from sub-standard sized specimens, expressed as energy absorbed at −80 °C, are summarised in Fig. 16.10. In the absence of added titanium (Fig. 16.10(a)), toughness was poor and little changed by thermal cycling or original grain size.

With 30 ppm Ti, the original fine grained weld metal was not as tough as the coarser deposit; this difference persisted after the inter-critically reheated (850 °C) and coarse-grained (1350 °C) simulations but disappeared after the two fine-grained cycles (1050 and 1150 °C), which generally gave the toughest weld metal. Intercritical reheating had no particularly detrimental influence on toughness.

The less tough deposits with ~100 ppm Ti gave poor toughness only in the as-welded and coarse-grained (1350 °C simulation) conditions. Other simulations, including intercritical reheating, were tougher, although not as tough as the deposits with 30 ppm Ti. The originally fine-grained weld metal remained less tough throughout all simulations.

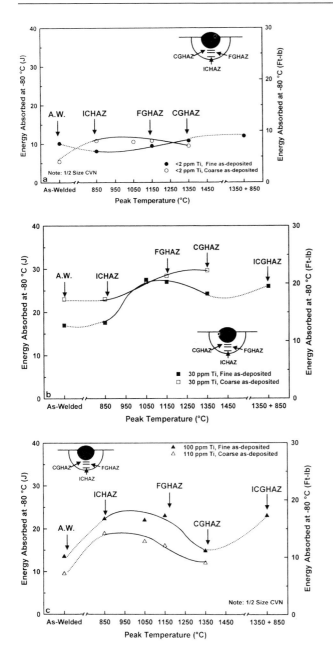

16.10 Effect of simulation treatment on Charpy energy of Mn–Ni–Mo weld metal with: *a*) <2 ppm Ti; *b*) 30 ppm Ti; *c*) ~100 ppm Ti; *d*) ~240 ppm Ti.

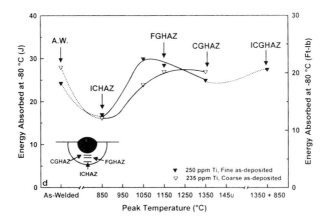

16.10 *(cont.)*

Only the ~240 ppm Ti weld showed a deterioration in toughness after the intercritical (850 °C) simulation, and that did not persist when it was preceded by a coarse-grained (1350 °C) cycle. Although toughness levels after other simulations were good, the results confirm the optimum titanium content of 30 ppm.

Discussion

The results show that, perhaps surprisingly, the influence of titanium on microstructure and toughness of complex Mn–Ni–Mo weld metal was very similar to its effects on the simpler C–1.4%Mn composition. As titanium was added, an optimum addition of approximately 40 ppm was followed by a deterioration and then a further improvement as Ti content increased to about 250 ppm. Nevertheless, the alloyed weld metal showed some distinctive features. The optimum interpass temperature was around 150 °C; toughness deteriorated if lower or higher interpass temperatures were used and strength fell appreciably with very high interpass temperatures.

Simulation showed that with both the favourable levels of titanium, the intercritically reheated microstructure gave most cause for concern, particularly with the higher favourable titanium level of 250 ppm.

References

1 Johnson MQ: 'Microstructure-property relationships in titanium-bearing high strength multipass shielded metal arc welds'. PhD thesis, Colorado School of Mines, Golden, Co, 1996.

2 Johnson MQ, Evans GM and Edwards GR: 'The influence of titanium additions and interpass temperature on the microstructures and mechanical properties of high strength SMA weld metals'. *ISIJ Int* 1995 **35**(10) 1222–31.
3 Johnson MQ, Evans GM and Edwards GR: 'The effect of thermal cycles on high strength steel SMA weld metal microstructures and properties'. Proc 4th int conf on 'Trends in welding research', Gatlinburg, USA, June 1995.

Part VI

Metallography

17

Metallographic features

This chapter describes some of the more complex metallographic observations from the previous investigations. The main topics are solidification features – segregation and non-metallic inclusions – and fine scale metallographic phenomena – dislocations and precipitation.

Segregation

In general, the weld metals examined appeared to be homogeneous with respect to most elements, except on a coarse micro-scale, i.e. within the distance of a prior austenite grain. However, obvious instances of banding and other compositional differences were noted in some deposits and those examined are the subject of the present section. No segregation was detected in C–Mn weld metals of normal impurity content using conventional light microscopy. Segregation was detected with high phosphorus contents and in some alloyed weld metals, including some Mn–Ni–Mo weld metals of high titanium content (Chapter 16). In weld metals containing nickel, this element increased the tendency for banding at weld ripple solidification fronts; this tendency was increased by manganese. Chromium gave a tendency for so-called white band formation.[1]

To examine the segregation behaviour of weld metals containing phosphorus and also nickel, investigations (Ref. 7, Ch. 6 and Ref. 2 and 3 this chapter) were carried out with the collaboration of the EO Paton Electric Welding Institute, who used techniques of micro-analysis after the microstructure had been revealed by etching in nital and sometimes in sodium picrate. The analysis probe had a diameter of 3 µm and thus analysed a region ~5 µm in diameter, although smaller inclusions having compositions greatly different from the matrix could readily be detected and sometimes identified by this method. Normally a mapping technique was used, scanning across the feature of interest which had previously been identified by etching and then lightly re-polished after marking the area.

Phosphorus

Welds deposited with three different heat inputs (1–4.3 kJ/mm), as described in Chapter 6, were used for these tests; attention was concentrated on deposits of high phosphorus content (0.04%). Microanalysis was carried out in such a way as to examine segregation which was likely at the solidification dendrite boundaries (Ref. 7, Ch. 6).

Even with the lowest heat input of 1 kJ/mm, dendrite boundary regions were found to be enriched in phosphorus, as well as in manganese and silicon. With increasing heat input the width of the segregated boundary region and the degree of phosphorus segregation increased. However, the liquation coefficients (the ratio of the maximum element concentration at the dendrite boundary to its maximum concentration at the dendrite axis) did not alter with heat input.

Microanalysis was also used to determine the composition of primary ferrite and of the second constituent, i.e. acicular ferrite or ferrite with second phase. The results of such tests, detailed in Table 17.1, revealed that carbon was much higher in the second constituent than in the ferrite but that manganese and silicon were the same. In contrast, phosphorus was increased in the second constituent at the higher heat input but not at the lower. It was concluded that the higher the heat input, the longer was the time available for segregation to proceed and, being more soluble in austenite than in ferrite, phosphorus enriched the last weld metal to solidify at the dendrite boundary regions.

Carbon and manganese

Segregation behaviour of unalloyed C–Mn weld metal was studied[2] to understand the behaviour of alloyed systems, particularly those containing nickel. The welds examined included those of the C–Mn system described in the second part of Chapter 3, deposited with a heat input of 1 kJ/mm. The concentration maps developed for welds of constant manganese content (1.8%) showed that heterogeneity of manganese increased considerably as the carbon content of the weld metal increased from 0.045 to 0.145%, as shown in Table 17.2. Increasing the manganese content at the highest carbon level also increased the manganese content of the second constituent in relation to the concentration in primary ferrite.

Table 17.1 Composition of microstructural constituents at different heat inputs

Heat input, kJ/mm	Micro-constituent	Element, wt%			
		C	Mn	Si	P
2.2	(Bulk analysis)	0.053	1.17	0.31	0.040
	Primary ferrite	0.07	1.13	0.29	0.040
	Second constituent	1.00	1.14	0.30	0.038
4.3	(Bulk analysis)	0.052	1.13	0.27	0.049
	Primary ferrite	0.038	1.12	0.28	0.038
	Second constituent	0.9	1.11	0.30	0.061

Table 17.2 Effect of carbon on composition of microstructural constituents

Bulk analysis, wt%		Mn, wt%, in:	
C	Mn	Primary ferrite	Second constituent
0.045	1.8	1.77	1.74
0.065	1.8	1.85	1.80
0.095	1.8	1.64	1.91
0.145	1.8	1.75	2.10
0.145	0.6	0.65	0.76
0.145	1.0	1.02	1.12
0.145	1.4	1.41	1.72
0.145	1.8	1.75	2.10

Etching in sodium picrate showed that particle density in the cell boundary regions of the second constituent increased with increasing carbon content.

Nickel and manganese

The welds examined were those containing up to 2.3%Ni described in Chapter 8, the compositions of which are given in Table 8.5. Etching in sodium picrate only slightly revealed the dendrite pattern,[3] probably because of the low contents of carbon, phosphorus and silicon – the elements which segregate most strongly – and the low heat input of 1 kJ/mm. Microanalysis, using the technique of concentration mapping, showed no difference between the concentration of nickel, manganese and silicon in grain boundary ferrite, acicular ferrite, and ferrite with aligned second phase, although the variability of analytical results was the greatest in acicular ferrite and the

Table 17.3 Chemical micro-heterogeneity of 2.25%Ni welds of varying manganese content

Bulk Mn content, wt%	Mn liquation coefficient	Difference in Mn content between cell axis and boundary
0.6	2.0	0.36
1.0	1.9	0.54
1.4	1.9	0.90
1.8	1.8	1.02

likelihood of differences in carbon content was noted (as was the case in the previous section).

Inclusions appeared to be predominantly of the manganese silicate type, with some manganese sulphide. The number of inclusions decreased with increasing manganese and nickel content. Although oxygen analyses were not undertaken, this decrease was likely to have been due to a fall in oxygen with increasing manganese, as noted in Table 3.1 in Chapter 3, rather than to any change in sulphur content.

As the nickel content was increased, the heterogeneity of nickel, manganese and silicon together increased. The increase in heterogeneity above 1%Ni would mean that in a 2.25%Ni–1%Mn weld metal, for example, some regions would contain as much as 2%Mn with between 1.7 and 2.8%Ni. This is illustrated in the results from Ref. 3, reproduced in Table 17.3. It was also noted that, despite the unchanging manganese liquation coefficient, the amount of segregation increased in terms of changes of composition, although the differences are roughly in the same ratio as the bulk manganese content.

These results may help to explain the lower toughness of more highly alloyed Ni–Mn weld metals. It was concluded that toughness depends predominantly on grain size up to 1%Mn, and above 1%Mn on heterogeneity.

Non-metallic inclusions

Inclusions are the products of deoxidation and desulphurisation reactions which have not been removed into the welding slag. Oxides formed at very high temperatures in the molten weld pool tend to be trapped in and removed by the slag; for this reason it is unusual, for example, to find weld metal inclusions containing magnesium. Evidence for the loss of inclusions during cooling before solidification also comes from results on submerged-arc welds of very high heat input where the weld oxygen content has been found to be lower than expected for the flux which had been used.[4] Oxygen content is a good measure of the volume fraction of oxide inclusions present in a weld metal.

The total inclusion content can be estimated from the bulk analysis as follows (Ref. 4, Ch. 1 and Ref. 4, Ch. 6). Assume that the oxygen in the weld metal first combines with Al to form Al_2O_3, then with Ti to form TiO and finally with Mn and Si to form $2MnO.SiO_2$; sulphur forms MnS. The multiplying factors to convert from wt% element to vol% inclusions (assuming the density of inclusions to be roughly half that of steel) are 4.09 for Al, 3.67 for Ti, 5.3 for S to MnS and 6.2 for any remaining O to $2MnO.SiO_2$*; 1% oxygen is equivalent to 1.125%Al and to 3%Ti. An alternative, simpler scheme,[5] assumes that the solubility of sulphur is the same in weld metals as in solid steel, i.e. 0.003%, and gives:

$$V_I = 10^{-2}\left\{5(O) + 5.4(S - 0.003)\right\} \qquad [17.1]$$

where V_I is the volume fraction of inclusions and (O) and (S) the weight percentages of oxygen and sulphur, respectively.

Although weld metal inclusions were originally thought to be glassy manganese silicates, manganese sulphides or mixtures of the two, electron microscopy and microanalysis have revealed complex structures, particularly at the surfaces of inclusions where, in the solid state, nucleation of acicular ferrite is thought to occur.

In several of the individual investigations comprising the whole research, studies of the details of inclusions were carried out by co-operating organisations, using methods which generally involved electron microscopy and microanalysis, sometimes with electron diffraction. The original papers should be consulted for details of these techniques.

The Mn–S system

Tests were carried out at Sheffield University, using techniques described in Ref. 6 and 7, on welds of the highest and lowest sulphur content of this system; details of these welds are given in Chapter 6 (bulk chemical compositions in Table 6.3). Average analyses of the inclusions, based on about 20 particles, are given in Table 17.4.

---

* As will be seen later in this Chapter, the manganese silicate appears to vary between $2MnO.SiO_2$ and $MnO.SiO_2$.

Table 17.4 Mean composition of non-metallic inclusions
in welds having varying sulphur content

Component, wt%	Weld sulphur content, wt%	
	0.007	0.046
SiO_2	52.9	43.8
MnO	39.1	45.4
TiO	5.4	5.5
S	1.2	4.5
Cu	1.3	0.8
Al_2O_3	<1.0	<1.0

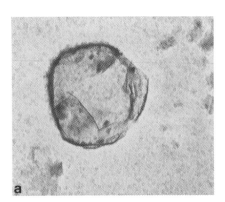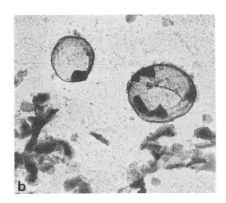

17.1 Microstructure of non-metallic inclusions in deposits with: *a*) 0.007%S; *b*) 0.046%S.

The inclusions were compounds of MnO and SiO_2 with smaller amounts of TiO, S and Cu. Increasing the sulphur content reduced the SiO_2 : MnO ratio from 1.35 to 0.95.

More detailed examination of the inclusions (Fig. 17.1) showed that they had surface coatings which tended to detach from the inclusions. In the low sulphur deposits these surface layers contained much titanium and had a face-centred cubic (FCC, a = 0.42 nm) structure, consistent with TiO. Some regions rich in sulphur and copper were also found, presumably a copper sulphide.

The inclusions in the high sulphur weld were more rounded, with a surface layer of MnS. Embedded in this surface layer were small particles (<100 nm) which were an almost stoichiometric TiO.MnO compound having a FCC lattice (a = 0.42 nm) and probably a spinel.

Identification of MnO, rather than MnO_2 as a major component of the inclusions is consistent with the discussion in Chapter 6 of the significance of Eq. [6.1] and [6.1a].

It has long been known[8] that non-metallic inclusions in basic weld metal have complex, multi-phase structures and that MnS exists as a separate phase. Of particular interest in the present work was the discovery of particles of a TiO/MnO compound embedded in the surface layer of inclusions in high sulphur weld metal. This change in composition of the surface layers from being rich in TiO to a more complex structure embedded in an MnS layer is likely to be responsible for the reduction in the proportion of acicular ferrite as weld sulphur content was increased (Fig. 6.13).

The results are similar to those of Devillers et al,[9] who found that aluminium manganese silicate inclusions were efficient at nucleating acicular ferrite in submerged-arc welds but that a covering of sulphur on the inclusions made such sites ineffective. The authors also suggested several possible mechanisms to explain the ease of acicular ferrite nucleation around inclusions:

(1)   compositional variations in the matrix around the inclusions (not substantiated by subsequent work);
(2)   relatively slight lattice disregistry between the crystals of ferrite and TiO;[10]
(3)   matrix straining around inclusions because of differences in coefficients of thermal expansion (this could partially account for 'poisoning' by MnS, the expansion coefficient of which is greater than that of ferrite, whereas other inclusions have lower coefficients).

The influence of copper

Investigations (Ref. 7, Ch. 8) carried out at the Institute of Materials Research, GKSS-Research Centre, on weld metals alloyed with 0–1.4%Cu and containing 1.4%Mn were described in Chapter 8 (the bulk chemical analyses being given in Table 8.7). Mean inclusion volume fractions and size distributions were determined on unetched samples with an SEM-IPS image analysis system operated at 15 kV and a magnification of ×3700. Carbon extraction replicas were obtained from lightly etched polished surfaces and thin foils were prepared using standard techniques. Transmission electron microscope studies were performed on a Philips EM 400 (S)TEM at 120 kV which was equipped with EDS microanalysis facilities. Semi-quantitative analyses were performed, corrections being made with standard software and thin film corrections.

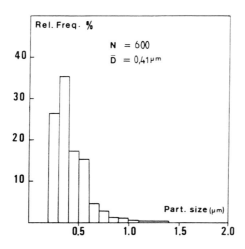

17.2 Size distribution of weld metal inclusions from typical welds containing copper.

The inclusion volume fraction and size distribution (Fig. 17.2) of the welds containing up to 1.4%Cu were constant, as would be expected from their similar oxygen and sulphur contents, averaging $0.040 \pm 0.002\%O$ and $0.007 \pm 0.001\%S$. The mean volume fraction was 0.2% from an analysis of ~600 particles in 50 fields. The values calculated from the relationships given at the beginning of the chapter are 0.29% for the more complicated relation and 0.21% using Eq. 17.1.

The inclusion size distribution, given in Fig. 17.2, shows that most inclusions had diameters between 200 and 600 nm, although a noticeable fraction (~1.2%) are between 1000 and 1400 nm (1 and 1.4 µm). The mean diameter was calculated to be 400 nm, from which the centre–centre inter-particle spacing was calculated to be 2.5 µm, assuming the inclusions to be spherical.

A typical inclusion is illustrated in Fig. 17.3, together with its X-ray elemental maps. The core appears to be a manganese silicate, as the ratio of Si : Mn of 0.56 obtained by semi-quantitative analysis is not very different from the expected value of 0.52 for $MnO.SiO_2$. Selected area diffraction (SAD) patterns, obtained from an inclusion using different beam directions, gave lattice parameters ($a = 1.172$, $c = 1.162$ nm) corresponding to the tetragonal oxide $MnO.SiO_2$, although patterns from other inclusions gave interplanar spacings closer to the monoclinic or triclinic $MnO.SiO_2$ phase.

Microanalysis, using the STEM mode with a small spot size (~10 nm), showed a titanium compound and copper or manganese

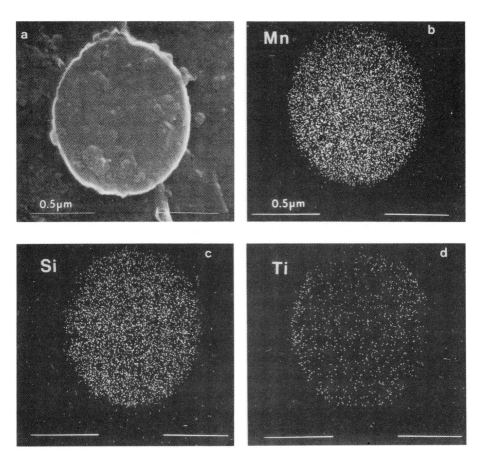

17.3 Secondary electron micrograph of an inclusion (*a*), together with its X-ray maps for Mn, Si and Ti (*b–d*).

sulphides at the edges. Where both were present, titanium was low in the sulphur-rich regions and vice-versa. Convergent beam diffraction (CBD) patterns from titanium-rich edges correspond to a cubic compound with a = 0.42 nm. This is consistent with TiO, TiC or TiN, which are all isomorphous and mutually soluble and have similar parameters. It would, therefore, be idle to speculate which compound is present without further evidence.

The CBD patterns from regions rich in copper and sulphur (which were low (~2%) in manganese) also had a cubic structure with a = 0.536 nm: this corresponds to $Cu_{1.6}S$, for which a = 0.541. From regions rich in manganese and sulphur, the CBD patterns were again cubic, but with a = 0.518 nm, which was attributed to α-MnS. The composition of the inclusions was not changed by varying the copper content.

Weld metals with aluminium

The weld metals examined (Ref. 1, Ch. 13) were from the first series examined in Chapter 12, with 1.4%Mn and ~40 ppm Ti; their compositions are those in Table 12.3 with 60 ppm N. The inclusions were examined by SINTEF in Norway,[11] using a SEM-based automatic image analyser to obtain compositional details and three-dimensional particle diameters.

The analytical results are plotted against aluminium content in Fig. 17.4 and 17.5, where they are superimposed on a ternary diagram showing the relevant oxides. In the absence of aluminium, most of the inclusion content corresponded to a manganese silicate containing appreciably more silica than is present in rhodonite ($MnO.SiO_2$), although a few pure silica inclusions were also present. The manganese silicate fraction of the inclusions remained on or close to a line along which the $SiO_2 : MnO$ ratio was ~1.64, regardless of the weld aluminium content. Added aluminium increased the alumina content of the inclusions until it was over 80% in the weld with 0.061%Al. A small amount of Ti-, Cu- and S-bearing material was also present; the last named was reduced with increasing aluminium content, as the bulk sulphur content fell from 0.007 to 0.004% (Table 12.3).

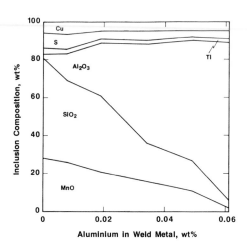

17.4 Influence of weld aluminium content on inclusion composition of weld metal with 1.4%Mn and 40 ppm Ti.

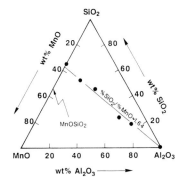

17.5 Weld metal inclusion compositions from Fig. 17.4 superimposed on ternary oxide system.

Histograms of inclusion size distribution in Fig. 17.6 show a mean 'three-dimensional size' of ~0.35 μm, which was independent of weld aluminium content. The maximum size rarely exceeded 1.0 μm and was never >1.7 μm. The average aspect ratio was approximately 1.2.

Examination of inclusions by STEM showed that those with a high alumina content rarely appeared to have nucleated acicular ferrite. High alumina inclusions, such as that shown in Fig. 17.7(*a*), showed evidence of some Mn, Cu and S at their surfaces but little trace of Ti. At low alumina contents, e.g. Fig. 17.7(*b*), the inclusion was of manganese silicate with surface compounds containing Mn, Cu, S and, sporadically, high levels of Ti.

Simple additions of titanium

The deposits examined were those described in Chapter 10 (Table 10.3), containing 1.4%Mn with additions of titanium giving weld metal titanium content up to 250 ppm. Above ~80 ppm Ti, oxygen content was noticeably lower and Mn and Si higher than the base composition. The lower oxygen led to a reduction in volume fraction of inclusions, as work on a series of alloyed weld metals with similar titanium additions to C–1.4%Mn weld metal (Ref. 1, Ch. 16) indicates (Fig. 17.8). Selected welds were examined at Sheffield University to determine inclusion compositions, using techniques described elsewhere.[6,7]

Results, plotted as for the aluminium series, are summarised in Fig. 17.9. In the absence of titanium, inclusions were of manganese silicate with small amounts of Cu and S; the ratio $SiO_2 : MnO$ was ~1.

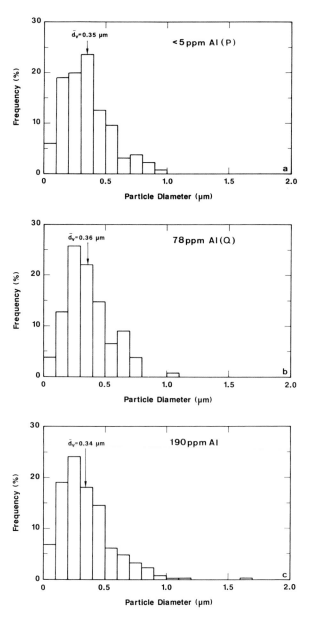

17.6 Size distribution of weld metal inclusions, each based on 500 measurements, from welds of varying aluminium content.

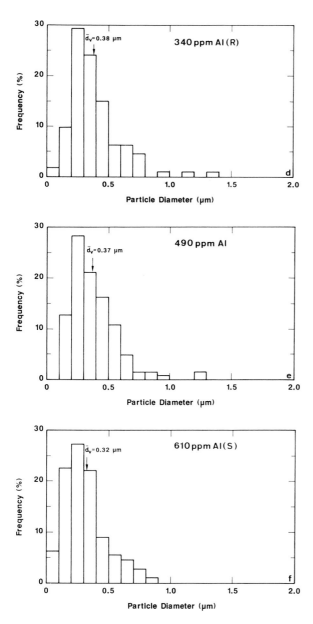

17.6 *(cont.)*

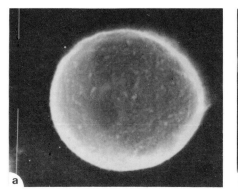 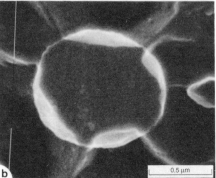

17.7 STEM micrographs of typical inclusions from welds of: *a*) High; *b*) Low aluminium content.

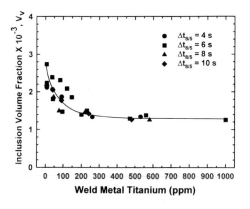

17.8 Influence of weld titanium content on inclusion content (calculated using Eq. 17.1) of low alloy weld metal whose compositions are given in Tables 16.1 and 16.2.

With increasing titanium, titanium oxide, TiO, was found and at the highest titanium content it formed almost 20% of the inclusions. Most of the reduction resulting from the increase in content of titanium was in silica, so that the SiO_2 : MnO ratio fell to ~0.8. The TiO content of the inclusions found in weld 'Q', with 28 ppm Ti (which showed the highest acicular ferrite content of the lower titanium welds) was 2.4%, a value close to that of 2% which has been claimed to be the optimum for submerged arc welds.[6]

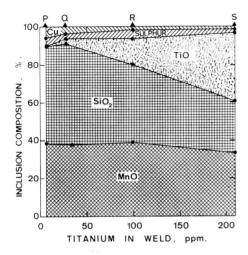

17.9 Influence of weld titanium content on inclusion composition of weld metal with 1.4%Mn.

Titanium with aluminium

In the 1.4%Mn welds with additions of both titanium and aluminium (Table 14.1), the weld with the highest content of both elements (500 ppm Al, 450 ppm Ti) contained cubic particles, which were shown in Fig. 14.3(*d*). Examination of replicas by TEM (Fig. 14.4) revealed both single and multiple cubic structures. Analysis by EDAX showed them to be complex Ti–Al compounds – probably oxides – with a wide range of compositions, ranging from titanium-rich (~65%Ti) with roughly 25%Al to aluminium-rich (60–80%Al) with 10–30%Ti, as shown in Table 17.5. Content of the elements Si, Mn and Fe was usually <10%.

Titanium with nitrogen – stereology

The welds examined were from boron-free deposits of C–1.4%Mn with additions of titanium and nitrogen, the composition of which is given in Table 12.1. The work described[12] was carried out at the École Polytechnique, Montréal, and is being continued to examine in detail the chemical composition of non-metallic inclusions in the welds.

The welds selected for examination were those of particular significance in Fig. 12.3(*a*) – a plot showing the dependence of the 100 J Charpy temperature on titanium at three nitrogen levels. Those selected were:

Table 17.5 Semi-quantitative analysis of inclusions in
welds with 1.4%Mn, 500 ppm Al and 450 ppm Ti

Element, wt%

Al	Ti	Si	Mn	Fe
21.7	65.3	13.0	n.d.	n.d.
27.3	64.3	1.7	4.9	1.8
60.8	30.2	0.9	7.0	1.1
62.9	26.4	n.d.	5.6	5.1
70.4	14.2	1.6	9.0	4.8
70.9	25.2	1.4	1.0	1.5
80.5	10.6	n.d.	7.9	1.0

Note: compositions calculated assuming no oxygen or
other light elements present.

~5 ppm Ti and poor toughness;
~35 ppm Ti and very good toughness;
~120 ppm Ti and relatively poor toughness;
~400 ppm Ti – again with good toughness.

Metallographically, the samples of the lowest nitrogen content contained approximately 12, 73, 60 and 65% acicular ferrite, respectively, for the four titanium levels examined.

The inclusion parameters, mean diameter, volume fraction and surface density, were measured on 100 randomly selected fields for each specimen, using back-scattered electron images at a magnification of 5000× in an electron microscope with an image analysing system. The results are summarised in Table 17.6.

At each nitrogen level, the inclusion volume fraction (Fig. 17.10), the number of inclusions analysed (Fig. 17.11) and the surface density of inclusions reached a peak at a titanium content of approximately 30 ppm, whilst the height of each peak increased as the nitrogen content increased. The decrease in these parameters as titanium increased above 30 ppm is a direct result of reduction in oxygen content as titanium is increased, as discussed in Chapter 10. Because the diameter of inclusions did not appear to vary significantly, and because the oxygen, sulphur and nitrogen contents did not change significantly between the titanium-free welds in each series and those containing ~30 ppm Ti, the increase in volume fraction and the other two parameters must have resulted from a change in composition of the inclusions and/or in their specific gravity.

Table 17.6 Results of stereological analysis of inclusions in C–1.4%Mn–Ti–N welds

Ti, ppm	N, ppm	Volume fraction, V_F, %	Mean circular diameter, d_M, μm	No. of inclusions	Surface density, S_V, cm⁻¹
1	80	0.22	0.34	960	28
28		0.23	0.32	1090	30
120		0.20	0.35	770	21
410		0.16	0.31	760	14
<5	160	0.25	0.35	1000	30
31		0.28	0.35	1140	34
51		0.28	0.38	970	31
120		0.26	0.37	960	30
410		0.16	0.36	560	15
<5	240	0.26	0.33	1240	34
29		0.31	0.34	1300	38
46		0.28	0.33	1260	35
120		0.19	0.35	760	22
450		0.17	0.33	750	20

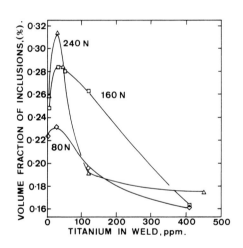

17.10 Influence of weld titanium content on volume fraction of inclusions in C–Mn–Ti deposits with varying nitrogen content.

High strength weld metals

Selected high strength welds of the Mn–Ni–Mo type with titanium contents up to 0.1% from Chapter 16 were subjected to analysis of inclusion size distribution using TEM examination of replicas at the

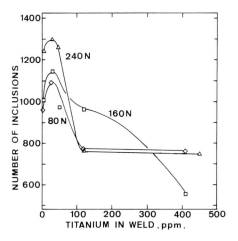

17.11 Influence of weld titanium content on number of inclusions analysed in C–Mn–Ti deposits with varying nitrogen content.

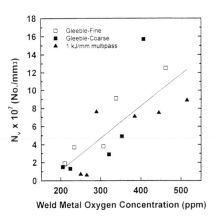

17.12 Influence of weld oxygen content on inclusion density.

Colorado School of Mines (Ref. 1, Ch. 16). The samples were extracted from multipass welds and simulated specimens. Their compositions are given in Tables 16.1 and 16.2 and the parameters measured are summarised in Table 17.7.

Because increasing titanium content reduced oxygen content, titanium reduced both inclusion volume fraction and also inclusion density (the number of inclusions per unit volume) (Fig. 17.12).

Table 17.7 Results of inclusion analysis on samples from 1.4%Mn–3.5%Ni–0.5%Mo welds with up to 1000 ppm Ti

Sample	Ti, ppm	O, ppm	d_v, μm	V_v, ×10⁻³	N_v ×10⁷ No./mm³	N_a ×10⁴ No./mm³	S_l, mm²/mm³	λ_v, μm	γ_g, μm	S_B, mm²/mm³	d_z, μm
F5	5	460	0.34	2.68	12.5	4.34	46.7	1.1	87	18.9	0.35
F30	30	340	0.28	1.07	9.1	2.55	22.6	1.2	91	17.3	0.15
F100	100	310	0.37	0.98	3.8	1.38	16.0	1.7	94	16.7	0.14
F250	250	230	0.40	1.20	3.7	1.46	18.2	1.7	104	15.1	0.19
F420	420	210	0.40	0.65	1.9	0.77	9.7	2	105	15.0	0.10
C5	5	410	0.31	2.42	15.7	4.84	47.0	1	91	17.3	0.40
C30	30	350	0.36	1.21	4.9	1.76	20.0	1.9	93	16.9	0.19
C110	110	320	0.38	0.81	2.9	1.76	13.4	1.8	95	16.5	0.12
C235	235	220	0.40	0.46	1.3	5.37	6.8	2.3	103	15.2	0.07
C340	340	200	0.41	0.52	1.5	5.96	7.7	2.2	105	15.0	0.07
E1	6	520	0.32	1.57	8.9	2.88	29.2	1.2	73	21.5	0.30
F1	38	440	0.39	2.35	7.5	2.93	36.0	1.2	78	20.1	0.28
H1	120	380	0.33	1.20	7.1	2.34	24.4	1.3	83	18.9	0.26
K1	230	290	0.36	1.80	7.6	2.71	30.3	1.3	93	16.9	0.21
L1	560	260	0.37	1.62	0.6	2.22	26.0	1.4	95	16.5	0.20
M1	1000	250	0.35	1.51	0.7	2.4	26.1	1.4	98	16.0	0.19

Note: d_v – mean inclusion diameter; V_v – inclusion volume fraction; N_v – number of inclusions per unit volume; N_a – number of inclusions per unit area; S_l – total inclusion surface area per unit volume; λ_v – mean centre-to-centre volume spacing; γ_g – austenite grain size; S_B – austenite grain area per unit volume; d_z – Zener diameter for inclusions.

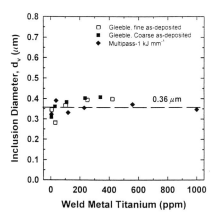

17.13 Influence of weld titanium content on inclusion diameter.

Despite these changes in inclusion volume fraction and density with titanium (and oxygen) content, inclusion diameter remained sensibly constant at 0.36 μm with a standard deviation of 0.037 μm (Fig. 17.13). In addition, size frequency distribution curves (which were log-normal) were similar for all welds examined. These findings are contrary to the view that inclusions act as inert substrates when nucleating acicular ferrite, a hypothesis which would require a sensible difference in inclusion diameter between welds in which acicular ferrite was nucleated and those in which it was not. However, the absence of detailed compositional data did not allow further ideas to be developed as to how acicular ferrite nucleation occurred.

Dislocations

The unusually high strength of weld metal in relation to its composition and grain size is a result of a very high dislocation density, brought about by cooling under high restraint (Ref. 1, Ch. 1). The network of dislocations is little affected by PWHT and is only destroyed by heating to normalising temperatures.

Dislocation densities were measured at the Welding Research Institute, Bratislava, Slovakia on selected high purity welds with and without additions of vanadium or niobium;[13] the composition of the deposits is given in Table 17.8. Measurements were made after stress relief for 2 hours at 580 °C, using the method of Ham.[14]

Typical dislocation substructures are shown in Fig. 17.14. The measured dislocation densities after PWHT were:

Table 17.8 Composition of weld metal having varying vanadium and niobium content

Element, wt%	C	Mn	Si	V	Nb	Ti	N	O
Designation								
52	0.077	1.33	0.31	0.0004	0.0006	0.0038	0.008	0.042
60	0.076	1.36	0.26	0.100	0.0005	0.0029	0.008	0.041
56	0.076	1.36	0.30	0.0005	0.094	0.0033	0.009	0.043

Note: all welds contained 0.006–0.008%S, 0.005–0.007%P, <5 ppm B, Al, ~0.03%Cr, Cu, Ni and ~0.005%Mo.

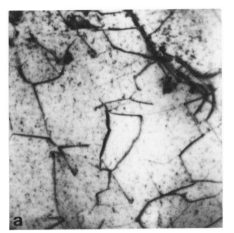

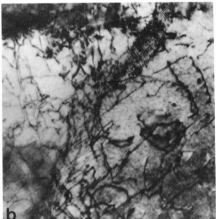

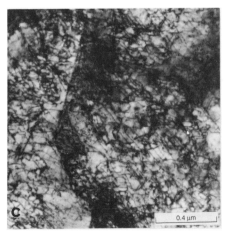

17.14 Arrangement of dislocations in stress-relieved 1.4%Mn weld metal with Ti and with and without V or Nb, showing: *a)* Low dislocation density without V or Nb; *b)* Extensive polygonisation with 0.1%V; *c)* Relatively high dislocation density with 0.094%Nb.

low Nb, low V $3.9 \times 10^9 \, \text{cm}^{-2}$
0.1%V, low Nb $8.0 \times 10^9 \, \text{cm}^{-2}$
0.094%Nb, low V $14.2 \times 10^9 \, \text{cm}^{-2}$

The dislocation density of the Nb-containing deposit was particularly high – probably as a result of precipitation during stress relief, as described in the next section.

These results show that the two microalloying elements each retarded elimination of dislocations during stress relief heat treatment, and that niobium was more effective on a weight percentage basis.

Precipitation/tempering

Stress relief and tempering after normalising of C–Mn weld metals led to precipitation of Fe_3C as described in Ref. 3, Ch. 13. This carbide was also identified in the films present in as-welded low C, low Mn deposits in Chapter 3.

Precipitation on stress relief – microalloyed welds

Microalloyed welds containing vanadium or niobium listed in Table 17.8 were examined for precipitation after PWHT at 580 °C, using TEM examination of thin foils.[13] No precipitates could be detected in the weld with vanadium but the weld with 0.094%Nb showed dispersed precipitates of niobium carbonitride of the NbX type (Fig.

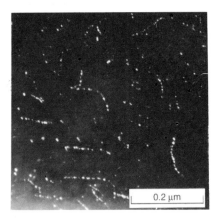

0.2 µm

17.15 Dark field micrograph of dispersed niobium carbonitride precipitates in 1.4%Mn weld with 0.094%Nb after stress relief at 580 °C.

17.15). The precipitates were in rows which decorated dislocations, thus inhibiting their movement, not only under plastic deformation but also when undergoing annihilation and polygonisation during heat treatment. This accounts for the high dislocation density measured after stress relief in comparison with the welds containing no microalloying elements except titanium or titanium with vanadium.

Precipitation in weld metal containing chromium

Tests using selected area diffraction and microchemical analysis (Ref. 3, Ch. 8) were used to examine precipitation after stress relief in selected weld metals containing chromium from those reported in Chapter 8, namely those with 1%Mn and 0.25 or 2.3%Cr; the bulk analyses of the welds are given in Table 8.1. With the lower chromium content, coarse precipitates were found (Fig. 8.5(a)), which diffraction showed to be cementite and microanalysis showed to contain some chromium, i.e. $(Fe,Cr)_3C$. Precipitates in the 2.3%Cr weld were much finer, as shown in Fig. 8.5(b) and were shown to be chromium nitride, CrN.

A larger investigation was therefore undertaken,[15] using weld metal of higher purity with 1%Mn, an addition of titanium and chromium contents up to 2.8%. The composition of the weld metal examined is given in Table 17.9.

Metallographic examination of the as-deposited regions showed that up to 1.5%Cr increased the proportion of acicular ferrite from just below 50% to ~70%, mainly at the expense of primary ferrite (Fig. 17.16). At higher chromium content primary ferrite continued to decrease, disappearing at 2.3%, whilst the proportion of acicular ferrite dropped sharply, both constituents being replaced by ferrite

Table 17.9 Composition of 1%Mn weld metal with chromium additions

Element, wt%

C	Mn	Si	S	P	Cr	Ti	N	O
0.068	1.06	0.35	0.006	0.009	0.027	0.0036	0.006	0.042
0.071	1.09	0.36	0.006	0.009	0.41	0.0041	0.007	0.042
0.072	1.05	0.35	0.005	0.007	0.79	0.0037	0.007	0.042
0.075	1.09	0.35	0.006	0.008	1.22	0.0052	0.009	0.042
0.070	1.06	0.34	0.006	0.008	1.57	0.0041	0.008	0.046
0.073	1.07	0.34	0.006	0.007	1.97	0.0040	0.009	0.043
0.077	1.05	0.35	0.006	0.008	2.28	0.0037	0.009	0.042
0.072	1.09	0.38	0.006	0.009	2.80	0.0033	0.008	0.047

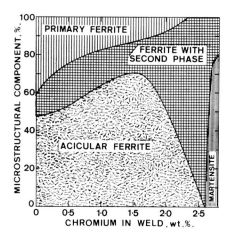

17.16 Influence of weld chromium content on as-deposited microstructure of 1%Mn welds with ~40 ppm Ti.

with second phase. At the highest chromium content of 2.8%, martensite, at 80%, had become the main constituent, the remainder being ferrite with second phase. The microphase between the laths of acicular ferrite was predominantly of the M/A (martensite/austenite) constituent at higher chromium levels, whilst at 2.8%Cr extensive carbide precipitation was found.

As chromium was added, the coarse-grained reheated weld metal was very close in appearance to as-deposited, as was noted in Chapter 8. The fine-grained region changed from a well defined equiaxed ferrite structure to colonies of ferrite with second phase and then to martensite, as chromium was increased. Above 1.6%, chromium greatly increased the proportion of carbide precipitates.

Precipitation in weld metals containing nickel or molybdenum

Stress relief at 580 °C of weld metal containing molybdenum (details of which are given in Chapter 8 and Table 8.3) (Ref. 4, Ch. 8) did not appear to change the microstructure seen by light or electron microscopy (Fig. 8.15). Precipitation in the intercritically reheated region (as-welded) (Fig. 8.16) was shown by diffraction to be cementite.

The only precipitation seen in Ni-bearing weld metal (Chapter 8, Table 8.5 and Ref. 6) was after stress relief heat treatment and was identified as carbide (presumably Fe_3C) resulting from decomposition of microphases.

Precipitation in copper-bearing weld metals

The weld metals examined at GKSS were from the standard compositions with 1.4%Mn with up to 1.4%Cu, which were described in Chapter 8, the bulk chemical analyses being given in Table 8.7. They had essentially acicular ferrite microstructure.

As-welded, precipitation of copper was evident in replicas from weld metals with 0.66 and 1.4%Cu (Ref. 7, Ch. 8), although such regions were small and seen only in limited areas; most areas contained no evidence of precipitation. In examination of thin foils, extensive copper precipitation was found in some areas of the 1.4%Cu deposit, as shown in Fig. 17.17 for top bead, unreheated weld metal. The intergranular precipitates were crystallographically related to the body-centered ferrite by the Kurdjumov/Sachs relation, in agreement with earlier work.[16,17] As can be seen from Fig. 17.17(a), the precipitates were randomly distributed and of different sizes. The larger particles (d ≤ 30 nm) were spheroidal in shape, the smaller (d ≤ 10 nm) were plate-like or needle shaped. In the regions free from copper precipitates, islands of twinned martensite were found (Fig. 17.18), whereas in regions containing precipitates, retained austenite was present (Fig. 17.18(c)).

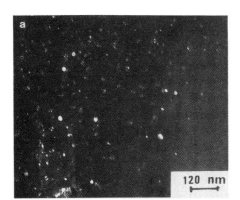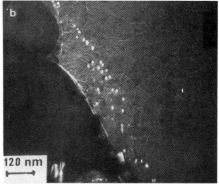

17.17 Dark field micrographs of ε–Cu precipitates in as-welded 1.5%Mn weld metal with 1.4%Cu showing: a) Intergranular precipitation; b) Precipitation adjacent to a grain boundary.

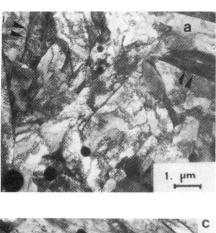

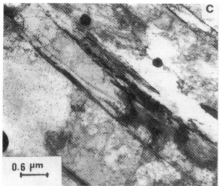

17.18 Bright field electron micrographs of as-welded 1.5%Mn weld metal with 1.4%Cu showing: *a*) Twinned martensite (arrowed) in a region free from copper precipitation; *b*) Twinned martensite at higher magnification; *c*) Retained austenite between ferrite laths in a region containing copper precipitation.

In some areas, rows of fine, closely spaced ε–Cu precipitates were found in the vicinity of ferrite grain boundaries (Fig. 17.17(*b*)), similar to interphase precipitation observed during isothermal transformation of Fe–Cu–Ni alloys.[17,18] This is consistent with growth of acicular ferrite by a ledge mechanism, at least in the later stages of the transformation, as described by Ricks et al.[19] In the precipitate-free zones of this weld extensive recovery was shown, as illustrated in Fig. 17.19. This twinned martensite, present only in the 1.4%Cu weld metal, no doubt contributed to its poor toughness in comparison with other copper-bearing weld metals.

Occurrence of isolated areas of copper precipitation in the as-welded state is hard to understand, unless areas locally high in copper were initially present. It has been shown that a cooling rate of 14 °C/sec is sufficient to retain copper in solid solution in a steel with 1.4%Cu,[20] although 285 °C/sec is required with 2.9%Cu. These cooling rates encompass the rates for the current welds, for which the

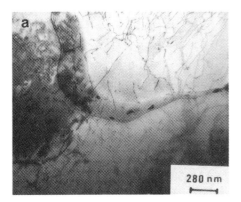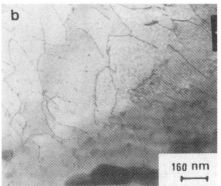

17.19 Bright field micrographs of a stress-relieved 1.5%Mn weld metal with 0.66%Cu showing: *a*) Fine copper precipitation associated with dislocations in a ferrite grain; *b*) Extensive recovery in a precipitate-free grain.

800–500 °C cooling time of 7 sec (Chapter 4) is equivalent to an average cooling rate between 800 and 500 °C of 43 °C/sec.

After stress relief, copper precipitation was not detected in weld metals containing less than 0.66%Cu. Fine precipitates associated with dislocations were observed in only a few isolated areas of the 0.66%Cu deposit.

Precipitation of ε–Cu was extensive in the 1.4%Cu deposit after stress relief. The intergranular precipitation was associated with dislocations, as reported by others,[17,18] the precipitates (Fig. 17.20(*a*)) being randomly distributed and spheroidal in shape. Extensive grain boundary precipitation was also observed, as in Fig. 17.20(*b* and *c*). Two types of this precipitation were seen; fine and closely spaced as in Fig. 17.20(*c*) and coarser as in Fig. 17.20(*b*). It is possible that the latter may have originally formed before stress relief and subsequently grown during heat treatment.

Discussion

The observations above can be summarised as follows. In welds without microalloying additions, non-metallic inclusions were of manganese sulphide and manganese silicate, in which the ratio of $SiO_2 : MnO$ could vary, somewhat unpredictably, between 0.5 and 1.6. Adding increasing amounts of aluminium replaced manganese silicate by alumina. At high aluminium and titanium levels, cubic

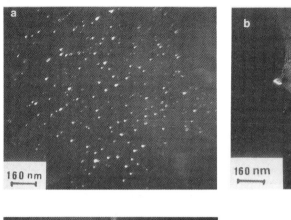

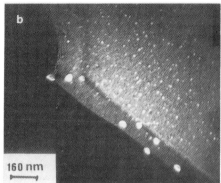

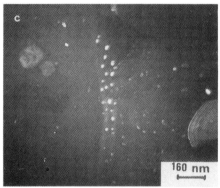

17.20 Dark field micrographs of ε–Cu precipitates in as-welded 1.5%Mn weld metal with 1.4%Cu showing: *a)* Intergranular precipitation; *b)* Fine intergranular and coarse grain boundary precipitates; *c)* Fine, closely spaced precipitates adjacent to a grain boundary.

inclusions, thought to be a mixture of Al_2O_3 and TiO, appeared. Mean inclusion diameter was within the range 0.34–0.4 μm, apparently independently of weld metal composition and heat input (1–2.2 kJ/mm).

When titanium was present, inclusion surfaces contained Ti, Mn, Cu and S. At high levels of sulphur, the inclusions were covered with MnS, in which were embedded Ti and Mn, probably as a TiO.MnO compound – possibly a spinel. At high aluminium levels, titanium was absent from inclusion surfaces. With high copper levels, copper sulphide (approximating to $Cu_{1.6}S$) appeared, as well as MnS. The titanium oxide appeared to be TiO, which is isomorphous with and mutually soluble in TiC and TiN. At the optimum titanium level for nucleating acicular ferrite, the titanium content of inclusions approximated to 2.4%.

Two of the numerous theories to account for nucleation of acicular ferrite have been discounted by the present work; namely compositional variations immediately around inclusions and inclusions acting as inert substrates. Nothing has been found in the present work to contradict the view that acicular ferrite requires manganese silicate inclusions with particles of a titanium compound – probably TiO – embedded in their surfaces and that this effect can be 'poisoned' by excessive sulphur or aluminium. The present results do not indicate an optimum composition of manganese silicate – in fact the ratio of SiO_2 to MnO in fairly similar welds has varied widely.

A further area of uncertainty is why, after a peak at ~30 ppm Ti, the efficiency of acicular ferrite nucleation fell and then improved with a titanium content about 5–10 times as high. It is possible that a spinel between TiO and MnO forms in some circumstances and this compound has a different efficiency as a nucleant for acicular ferrite from that of TiO.

The role of aluminium is believed to be twofold: firstly it replaces manganese silicate by alumina and secondly, by using available oxygen preferentially, it lessens the amount of TiO which can be formed. This behaviour would make TiO a more likely nucleant for acicular ferrite than one of its isomorphs, TiC or TiN.

Unfortunately, the extremely low atomic weight of boron makes its detection difficult on a microscopic scale and, hence, the reasons for its behaviour must remain enigmatic until new micro-analytical techniques are developed for its detection and estimation.[21]

Nitrogen in inclusions was only found in sample X2, containing 400 ppm Ti and 240 ppm N. It is unlikely that nitrogen reacts with titanium to form TiN. Indeed, what seems to happen is that the nitrogen reacts with TiO to form Ti(O,N), or possibly Ti(C,O,N), as proposed by Tiersma et al.[22]

Dislocations, precipitates and 'second phases' are clearly important features of ferritic weld metals, as exemplified by weld C_0, the deposit free from microalloying elements. As-welded, this material gave 100 J at −15 °C, whereas after stress-relief at 580 °C for 2 hours, the same notch toughness was achieved at −70 °C. The top bead remained essentially free from acicular ferrite and the change, apart from a slight softening, was a breakdown of grain boundary carbide films and their spheroidisation. No direct link thus exists between the acicular ferrite content of the top bead and Charpy toughness near the centreline of a multirun deposit. Up to a point, however, trends could progress in parallel, as in the case of manganese, which progressively increased the proportion of acicular ferrite between 0.6 and 1.8%, whereas the optimum toughness was at 1.4%Mn.

References

1 Koromzay T: 'The effect of welding processes on ferrite band formation in stress-relieved low-alloyed weld metal'. IIW DocXII–E–24–81.

2 Pokhodnya IK, Voitkevitch VG, Shevchenko GA and Evans GM: 'Relationship between structure and chemical microheterogeneity of welds alloyed with nickel–manganese and carbon–manganese'. IIW DocII–A–822–90.

3 Pokhodnya IK, Voitkevich VG, Denisenko AV and Evans GM: 'Investigation into chemical heterogeneity of weld metal alloyed with Ni & Mn'. IIW DocII–A–751–88.

4 Bailey N: 'Submerged-arc welding ferritic steels with alloyed metal powders'. *Weld J* 1991 **70**(8) 187s–206s.

5 Kluken AO and Grong Ø: 'Mechanisms of inclusion formation in Al–Ti–Si–Mn deoxidised steel weld metals'. *Metall Trans A* 1989 **20A** 1335–49.

6 Saggese ME et al: 'Factors influencing inclusion chemistry and microstructure in submerged-arc welds'. Proc TWI conf 'The effects of residual, impurity and micro-alloying elements'. London, Nov 1983, Paper 15.

7 Bhatti AR et al: 'Analysis of submerged-arc welds in microalloyed steels'. *Weld J* 1984 **63**(7) 224s–30s.

8 Boniszewski T, le Dieu SE and Tremlett HF: 'Sulphur behaviour during deposition of mild steel weld metal; Part 1: Deposition of covered electrodes from various classes on high strength steel'. *Brit Weld J* 1966 **13** 558–77.

9 Devillers et al: 'The effect of low level concentrations of some elements on the toughness of submerged-arc welded C–Mn steels'. Proc TWI conf 'The effects of residual, impurity and micro-alloying elements', London, Nov 1983, Paper 1.

10 Mori N et al: 'Mechanism of notch toughness improvement in Ti–B bearing weld metals'. IIW DocIX–1196–81.

11 Kluken AO, Grong Ø and Hjelen J: 'SEM-based automatic analysis of non-metallic inclusions in steel weld metals'. *Mat Sci Tech* 1988 **4** 649–54.

12 Blais C and L'Esperance G: 'Stereological characterisation of inclusions found in 14 samples of the Ti–N series'. Interim report to Oerlikon Welding from CCMM, Montréal, Oct 1995.

13 Bosansky J and Evans GM: 'Relationships between the properties of weld metals microalloyed with V and Nb, their structure and properties'. *Welding Intl* 1992 **6** 997–1002.

14 Ham RK: 'The determination of dislocation densities in thin films'. *Phil Mag* 1961 (6) 1183.

15 Jorge JCF, Rebello JMA and Evans GM: 'Microstructure and toughness relationships in C–Mn–Cr all-weld metal deposits'. IIW DocII–A–880–93.

16 Speich GR and Oriani RA: 'The rate of coarsening of copper precipitate in an alpha–iron matrix'. *Trans Met Soc AIME* 1965 **233** 623.

17 Howell PR, Ricks RA and Honeycombe RWK: 'The observation of inter-phase precipitation in association with the lateral growth of Widmanstät-ten ferrite'. *J Mat Sci* 1980 **15** 376–80.

18 Ricks RA, Howell PR and Honeycombe RWK: 'The nature of acicular ferrite in HSLA steel weld metals'. *Met Trans* 1979 **10A** 1049.

19 Ricks RA, Howell PR and Baritte GS: 'The effect of Ni on the decom-position of austenite in Fe–Cu alloys'. *J Mat Sci* 1982 **17** 732.

20 Wada H et al: 'Strengthening effects of copper in structural steels'. State University of Ghent, Belgium, June 1983.

21 Blais C et al: 'Characterisation of small inclusions'. Proc 'Microscopy and microanalysis 96', MSA/MAS and MSC/SMC, Minneapolis, USA, Aug 1996.

22 Tiersma T et al: 'Structure-property relationships in (low) C–Mn weld metals with varying amounts of boron and titanium'. Proc IWC–87, New Delhi, India, 1987.

18

Summary

Weld metal titanium content has been a recurrent theme of this work, following the finding that the influence of rutile and other compounds of titanium is totally different from that of other minerals in the electrode coating. Although the yield of titanium is low, the effect is reproducible, since additions of 5% rutile to the dry mix (giving about 40 ppm Ti in weld metal) can be accurately made, even under production conditions. To develop the maximum response, a manganese content of 1.4% in the weld metal was necessary to attain the required degree of hardenability in the plain C–Mn system. Furthermore, two optima in toughness were found, one at 30–40 ppm Ti and the other at 300–400 ppm Ti.

There was a negative aspect in that use of titanium metal powder in the coating, required to attain the higher level, resulted in an increase in weld metal hydrogen content; thus, the source of this raw material is important. Hydrogen has not been studied in this monograph but has naturally been a subject of parallel research.[1] For example, the tertiary binder, used to obtain extra low weld metal hydrogen levels, was developed during the initial stages of the programme and the technology was transferred to a Swiss silicate producer for commercialisation.

A manganese content of 1.4% was also found to be the most favourable to toughness over a range of carbon and silicon contents – which were optimised at 0.07%C and 0.2%Si – as well as after stress relieving and also after normalising and tempering. Both sulphur and phosphorus were confirmed as being harmful and should be maintained as low as practicable. Process variables such as electrode diameter, interpass temperature, heat input and welding position had essentially no effect on the optimum manganese level. This is fortunate from a practical standpoint, since a single consumable can be used to cover a range of applications.

Intermediate amounts of titanium, in the neighbourhood of

150 ppm, caused embrittlement of 1.4%Mn weld metal and this probably accounts for apparent anomalies in the published literature. For example, Taylor[2] reported that 1.7%Mn was required to achieve optimum toughness in weld metal deposited from E7016 electrodes.

In retrospect, earlier controversies[3,4] about the differences between the E7016 and E7018 types of electrode are now considered to be accounted for by the different deoxidants present in the coatings of the electrodes tested. Independent tests have shown little or no influence of the ratio of calcium carbonate to calcium fluoride, whereas addition of metallic magnesium, for example, can substantially reduce the content of oxygen, nitrogen and sulphur and can also increase recovery of titanium into the weld metal.

An essential requirement for any slag system is to ensure that weld metal composition is balanced, so as to produce optimum mechanical properties by generating acicular ferrite and the correct microphase morphology. With a reduction in weld metal oxygen level, less than 30 ppm Ti was necessary for peak properties, whereas with an increase in nitrogen content, approximately 50 ppm Ti was required for the same purpose.

Early in the programme, it was found that addition of the major alloying elements chromium, molybdenum and nickel caused peak properties to be displaced to lower manganese contents, indicating an interchangeability of these elements. A threshold was reached for chromium and molybdenum but nickel additions continued to be beneficial if manganese content was sufficiently reduced. For example, 3.5%Ni deposits were found to be optimised for toughness at 0.5%Mn. This knowledge was used to develop basic flux-coated electrodes for the offshore sector[5] and a full range of EXX18 and EXXX18 type consumables was made available.[6]

Subsequent work concentrated on individual effects of the microalloying elements titanium, aluminium, niobium, vanadium and boron on toughness at two distinct manganese levels. With 1.4%Mn, all trace elements were initially beneficial to some extent. Conversely, at 0.6%Mn, no essential benefit was derived. At both manganese levels, the presence of niobium and vanadium caused a poor response after stress relieving. Combinations of trace elements were then studied, at the higher of the two manganese levels, and interactive effects were quantified. In the main, aluminium tended to destroy the beneficial effect of titanium on toughness, whereas boron, at a specific concentration of ~40 ppm, tended to enhance it.

Addition of nitrogen to Ti–B and Ti–B–Al combinations proved to be particularly damaging and nitrogen control is an essential requisite for such types of weld metal. On the other hand, deposits con-

taining an excess of aluminium and boron were improved by intermediate amounts of nitrogen. It was found that nitrogen induced acicular ferrite formation under certain circumstances, and it is possible
that future developments using microalloying elements will rely on
ambient nitrogen to yield superior properties.

Effects of the major alloying elements, including copper, were re-
assessed at two levels of manganese, both in the presence and absence
of a small titanium addition. With 1.4%Mn, the response was mainly
negative, whereas at 0.6%Mn a marked initial improvement occurred,
especially on adding nickel in the presence of 30 ppm Ti. A comparison of the major alloying elements and the trace elements showed
opposing trends. For the former a low manganese content was beneficial, whereas for the latter the higher manganese level was a requirement. There is thus a complex situation with no single element being
universally advantageous. Furthermore, in the case of titanium, there
were twin peaks at 1.4%Mn (and other manganese contents), intermediate contents of titanium being particularly harmful when trace
amounts of vanadium were not present.

Extension of the work to high strength Mn–Ni–Mo deposits
showed, rather surprisingly, the universality of the titanium effects.
Twin optima in notch toughness were again encountered at 30 and
300 ppm Ti. The initial microstructure was bainitic and was totally
modified by titanium. The optimum interpass temperature was
150 °C, strength falling appreciably at high heat inputs and interpass
temperatures. In the main, the upper titanium peak appeared more
robust with regard to procedural variations than the lower.

Although a substantial data base has been built up, the mechanism
producing such a profound effect has not been identified, other than
being related in some way to the detailed structure of the surfaces of
non-metallic inclusions. In consequence, studies will continue on a
co-operative basis in a number of laboratories world-wide, so as to
establish exactly where titanium is located and how it so drastically
affects transformation products and their toughness. A proposal,
therefore, is to add increments of 5 ppm Ti progressively to the
microalloy-free deposit (C_0) and, by spanning the range up to
30 ppm, to study the changes causing the volume fraction of acicular
ferrite to increase from 10 to almost 80%. Subsequently, non-metallic inclusions are to be investigated for light elements, so as to separate the components making up the different Ti, Ti–B and Ti–B–Al
systems, at three levels of nitrogen.

In addition, transformation characteristics of the selected welds are
to be determined and further simulation tests carried out to enable
effects of thermal cycling to be evaluated. Weldments are also to be

subjected to strain ageing and, using a fracture mechanics approach, tested to assess in detail the relative susceptibility of different groups of weld metal to embrittlement. Finally, neural network analysis is to be applied to the complete data set in an attempt to make predictive assessments.

References

1 Evans GM and Baach H: 'The hydrogen content of welds deposited by different welding processes'. *Z fur Schweisstechnik/Journal de Soudre* 1975 **65**(4) 93–101.
2 Taylor, DS: 'The effect of manganese on the toughness of E7016 type weld metal'. *Weld Metal Fab* 1982 **50**(11), 452–60.
3 Boniszewski T: 'The effect of iron powder in basic low hydrogen all-positional electrodes'. Int conf on 'Trends in steels and consumables for welding', TWI, London, Nov 1978, Paper 15.
4 Evans GM: 'The effect of iron powder in basic low hydrogen all-positional electrodes'. *Schweissmitteilungen* 1982 **40**(100) 25–35.
5 Taylor DS and Evans GM: 'Development of MMA electrodes for offshore fabrication'. *Metal Construction* 1983 **15**(8) 438–43.
6 Evans GM: 'Basic low-alloy steel covered arc welding electrodes according to AWS A5.5–81'. *Schweissmitteilungen* 1987 **45**(113) 22–3.

Index

Note: the term 'pure welds' indicates weld metals with the elements Al, B, Nb, Ti, V each below 5 ppm, unless deliberately added. 'Standard welds' contain typically 120 ppm V, 55 ppm Ti, 20 ppm Nb and 5 ppm Al.